Lecture Notes on Coastal and Estuarine Studies

Managing Editors:
Richard T. Barber Christopher N. K. Mooers
Malcolm J. Bowman Bernt Zeitzschel

8

Marine Phytoplankton and Productivity

Proceedings of the invited lectures to a symposium
organized within the 5th conference of the European
Society for Comparative Physiology and Biochemistry –
Taormina, Sicily, Italy, September 5–8, 1983

Edited by O. Holm-Hansen, L. Bolis and R. Gilles

Springer-Verlag
Berlin Heidelberg New York Tokyo 1984

Managing Editors
Richard T. Barber
Coastal Upwelling Ecosystems Analysis
Duke University, Marine Laboratory
Beaufort, N.C. 28516, USA

Malcolm J. Bowman
Marine Sciences Research Center, State University of New York
Stony Brook, N.Y. 11794, USA

Christopher N. K. Mooers
Dept. of Oceanography, Naval Postgraduate School
Monterey, CA 93940, USA

Bernt Zeitzschel
Institut für Meereskunde der Universität Kiel
Düsternbrooker Weg 20, D-2300 Kiel, FRG

Contributing Editors
Ain Aitsam (Tallinn, USSR) · Larry Atkinson (Savannah, USA)
Robert C. Beardsley (Woods Hole, USA) · Tseng Cheng-Ken (Qingdao, PRC)
Keith R. Dyer (Taunton, UK) · Jon B. Hinwood (Melbourne, AUS)
Jörg Imberger (Western Australia, AUS) · Hideo Kawai (Kyoto, Japan)
Paul H. Le Blond (Vancouver, Canada) · Akira Okubo (Stony Brook, USA)
William S. Reebourgh (Fairbanks, USA) · David A. Ross (Woods Hole, USA)
S. Sethuraman (Raleigh, USA) · John H. Simpson (Gwynedd, UK)
Robert L. Smith (Corvallis, USA) · Mathias Tomczak (Cronulla, AUS)
Paul Tyler (Swansea, UK)

Editors
Scientific Editor
Dr. O. Holm-Hansen
Food Chain Research Group
Scripps Institution of Oceanography, University of California
La Jolla, CA 92093, USA

Prof. L. Bolis
Laboratory of General Physiology, University of Messina
Via dei Verdi 85, Messina, Italy

Prof. R. Gilles
Laboratory of Animal Physiology, University of Liège
22, Quai de Beneden, 4020 Liège, Belgium

ISBN 3-540-13333-X Springer-Verlag Berlin Heidelberg New York Tokyo
ISBN 0-387-13333-X Springer-Verlag New York Heidelberg Berlin Tokyo

This work is subject to copyright. All rights are reserved, whether the whole or part of the material is concerned, specifically those of translation, reprinting, re-use of illustrations, broadcasting, reproduction by photocopying machine or similar means, and storage in data banks. Under § 54 of the German Copyright Law where copies are made for other than private use, a fee is payable to "Verwertungsgesellschaft Wort", Munich.

© by Springer-Verlag Berlin Heidelberg 1984
Printed in Germany

Printing and binding: Beltz Offsetdruck, Hemsbach/Bergstr.
2131/3140-543210

EUROPEAN SOCIETY FOR COMPARATIVE PHYSIOLOGY AND BIOCHEMISTRY

5th Conference - Taormina, Sicily - Italy, September 5-8, 1983

This volume gathers the proceedings of the invited lectures of the ymposium on Marine Phytoplankton and Productivity.

Conference general theme and symposia

Physiological and Biochemical Aspects of Marine Biology

Symposia : 1) Toxins and drugs of marine animals
2) Responses of marine animals to pollutants
3) Marine phytoplankton and productivity
4) Osmoregulation in estuarine and marine animals

Conference Organization

General organizers

L.BOLIS and R.GILLES
Messina, Italy / Liège, Belgium

Symposium scientific organizer

for Marine phytoplankton and productivity
O.HOLM-HANSEN
La Jolla, California, USA

Local organizers

G.STAGNO d'ALCANTRES, S.GENOVESE, G.CUZZOCREA, F.FARANDA,
A.CAMBRIA
Messina, Italy

Local secretariat

A.SALLEO and P.CANCIGLIA
Messina, Italy

Conference under the patronage of

The University of Messina, Italy
The Fidia Research Laboratories, Italy
The European Society for Comparative Physiology and Biochemistry.

PREFACE

When I was asked to organize this symposium on marine productivity, it made me reflect on what aspects of this subject would be stimulating to a heterogeneous group of laboratory-oriented physiologists and biochemists. In recent years there have been several books which discusses the methodology commonly used in primary production studies and described the magnitude of photosynthetic CO_2 reduction in various areas of the world's oceans. I therefore decided to dispense with these conventional aspects of primary production and invite researchers to speak on a variety of problems relating the abundance and activity of phytoplankton to environmental conditions. The lectures I invited were thus quite diverse in character, but all were related either to factors affecting the rate of photosynthesis or to the fate of reduced carbon as it passes through the microbial food web.

In addition to these talks the participants benefited from a number of shorter presentations and poster sessions which dealt with production and cycling of organic carbon in the marine environment.

February 1984   Osmund HOLM-HANSEN

CONTENTS

1. Factors Governing Pelagic Production in Polar Oceans
   E.SAKSHAUG and O.HOLM-HANSEN 1
2. Productivity of Antarctic Waters. A Reappraisal
   S.Z. EL-SAYED 19
3. A Thermodynamic Description of Phytoplancton Growth
   D.A. KIEFER 35
4. Mechanisms of Organic Matter Utilization by Marine Bacterio-plankton
   F.AZAM and J.W. AMMERMAN 45
5. Phytoplankton Solved the Arsenate-Phosphate Problem
   A.A.BENSON 55
6. Excretion of Organic Carbon as Function of Nutrient Stress
   A.JENSEN 61
7. Seasonal Changes in Primary Production and Photoadaptation by the Reef-Building Coral *Acropora granulosa* on the Great Barrier Reef
   B.E.CHALKER, T.COX and W.C.DUNLAP 73
8. General Features of Phytoplankton Communities and Primary Production in the Gulf of Naples and Adjacent Waters
   D.MARINO, M.MODIGH and A.ZINGONE 89
9. Understanding Oligotrophic Oceans : Can the Eastern Mediterranean be a Useful Model ?
   T.BERMAN , J.AZOV and D.TOWNSEND 101
10. Growth Rates of Natural Populations of Marine Diatoms as Determined in Cage Culture
    G.A.VARGO 113
11. Observed Changes in Spectral Signatures of Natural Phytoplankton Populations : The Influence of Nutrient Availability
    C.S.YENTSCH and D.A.PHINNEY 129
12. Flow Cytometry and Cell Sorting : Problems and Promises for Biological Ocean Science Research
    C.M. YENTSCH, T.L.CUCCI and D.A.PHINNEY 141
13. Determination of Absorption and Fluorescence Excitation Spectra for Phytoplankton
    B.G.MITCHELL and D.A.KIEFER 157

SUBJECT INDEX 171

FACTORS GOVERNING PELAGIC PRODUCTION IN POLAR OCEANS

E. SAKSHAUG [1] and O. HOLM-HANSEN [2]

[1] University of Trondheim, Biological Station,
N-7000 Trondheim, Norway

[2] Polar Research Program, Scripps Institution of
Oceanography, University of California, La Jolla,
California 92093, USA.

INTRODUCTION

The standing stock of pelagic primary producers is conveniently expressed as the difference between growth and reduction of the standing stock by grazing or by sinking out of the euphotic zone. The rate of growth is determined largely by environmental variables such as the light regime, temperature, and nutrient supply. In polar regions these variables are generally quite different than those in lower latitudes. The atmospheric light regime, for instance fluctuates extremely through the season at high latitudes. Sea ice affects the submarine light regime profoundly, and the formation and melting of ice affect the stability of the water column.

The Arctic and Antarctic Oceans do, however, also differ in certain respects : size, topographical features, large-scale water transport patterns, and the magnitude of nutrient supply to the euphotic zone. Therefore we cannot assume *a priori* that the primary production in the two areas are governed by environmental variables in an identical fashion even if the seasonal variation in phytoplankton standing stocks may not be notably different.

In this paper we attempt to review existing knowledge regarding phytoplankton growth *vs*. environmental variables in polar seas in relation to geographical and seasonal distribution of standing stock. We shall deal only with pelagic primary producers, which account for most primary production in polar regions. Ice biota, of which the ecology has been reviewed elsewhere (Horner 1976, 1977), often attain a high standing stock but shows low production due to light limitation (Holm-Hansen and Huntley 1984). Benthic algae, although of much ecological importance in specific habitats, also contribute but a small percentage of the total primary production at high latitudes. As a "background" to viewing primary production in the polar regions, it is of interest to note that fisheries experts predict that the Antarctic seas can sustain a krill harvest of over 100 million metric tons a year, as compared to the present total world harvest of about 75 million tons per year of fish and shellfish (Gulland 1970). Does this reflect very high primary production in the Antarctic, or does it merely reflect a short and efficient food chain ?

PHYSICAL AND CHEMICAL FEATURES

The major difference between the Arctic and the Antarctic involves the relative positions and areas covered by land og sea (Fig. 1). The south polar region is covered by a huge (14×10^6 km^2) and high continent, which is surrounded by the unimpeded flow of the East and West Wind Drift systems. In sharp contrast, the north polar region is occupied by the Arctic Ocean basin, which is surrounded by large land masses which affect both water transport and meteorological conditions over the entire Arctic region.

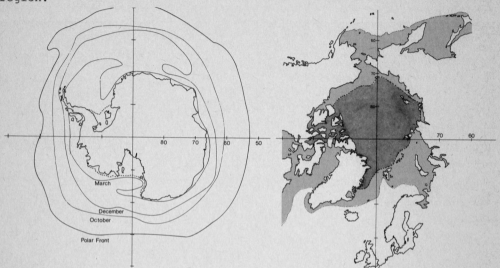

Fig. 1. Maps of the Antarctic Ocean (left), showing the Polar Front and the maximum and minimum extent of sea ice (after El-Sayed, 1970a), and the Arctic Ocean (right), showing maximum and minimum extent of sea ice (based on Polar Regions Atlas by CIA).

It is convenient to define the Antarctic marine ecosystem as those waters south of the Antarctic Polar Front, which is located between 48-60°S and generally close to about 55°. Most of the Antarctic Ocean is therefore situated at latitudes corresponding to the zone from northern France to northern Norway. There is no such clear definition of Arctic waters. One operational definition, as in this paper, is to consider as "Arctic" those waters which are covered by the maximal extent of sea ice. When defined in this manner, the Arctic marine environment stretches south to 58-63°N in the Bering and Greenland Seas. The North European region is anomalous due to the influx of warm Atlantic water. Thus the whole Norwegian coast and the Barents Sea south of 73°N and west of 35°E remain premanently ice free. In this part of the Arctic the Polar Front is situated as far north at 77 70° which is north of the ice edge in winter.

The areas covered by ice or water in both polar regions are shown in Table 1.

The two polar oceans are similar with respect to the prevalence of

	Antarctic Ocean	Arctic Ocean
area[1]	30	14
sea ice, maximum	18 (Oct.)	14 (March)
sea ice, minimum	2.5 (March)	7 (Aug.)
area with seasonal ice	15.5	7

Table 1. Area and distribution of sea ice (10^6 km^2) in the Antarctic and Arctic Oceans. Data from Jacka (1981) and Walsh and Johnson (1979). Total world ocean area is 361×10^6 km^2

[1]) Antarctic Ocean: south of Polar Front; Arctic Ocean: area of maximum ice cover.

ice and low sea temperatures, which range from about +5° at the Polar Front to -1.8° in ice-filled waters.

In the Arctic Ocean the large-scale water transport is characterized by an influx of Atlantic water west of Svalbard and to a lesser extent in the Barents Sea. Influx through the Bering Strait may be only 1/3rd of that from the Atlantic (Treshnikov and Baranov 1977). At the Polar Front Atlantic water is submerged, but remains distinguishable for vast areas at depths >100m. Outflow is mainly by the East Greenland and Labrador currents. This water is cold and has a low salinity (29-34°/oo) due to ice melting as well as discharge from Siberian and North American rivers. The ice of the central polar basin exhibits a slow anticyclonic drift.

In the Antarctic Ocean the West Wind Drift is the dominating feature. It is characterized by net upwelling, particularly at the Divergence, which separates the West Wind drift from the East Wind drift, and a northern component of surface layer flow. Downwelling prevails at the Polar Front and close to the continent where bottom water is formed, particularly in the Weddell Sea.

The continuous upwelling in the Atlantic Ocean results in the highest nutrient concentrations found anywhere in surface waters. A decrease is apparent during phytoplankton blooms, but complete exhaustion of surface nutrients is an unlikely event, at least in offshore waters. In contrast, Arctic surface waters exhibit maximum nutrient values close to those of the North Atlantic and nutrient exhaustion after a bloom. Table 2 shows that the minimum level for offshore Antarctic waters is actually higher than the maximum levels for the Arctic ocean. Another notable difference is the 8-fold higher level of silicate compared to nitrate and phosphate in the Antarctic relative to the Arctic Ocean. This is apparent also in the consumption pattern for blooms (assuming the difference between maximum and minimum levels to represent conversion into biomass), implying that bloom-forming diatoms in the Antarctic are silicified to an extent far beyond Arctic diatoms.

The extreme seasonal variation in the light regime is revealed in Fig. 2. The maxium sun height decreases with increasing latitude, for

	Barents Sea		Scotia Sea	
	before	after	before	after
NO_3	9-14	~0	32	12
NH_4	0-2	0-2	0-2	0-4
PO_4	0.5-0.6	~0.1	1.9	0.9
SiO_2	4-5	~0	102	44
NO_3/PO_4	17-23	17-23 (Δ)	17	20 (Δ)
NO_3/SiO_2	25-28	2.0-2.8 (Δ)	0.31	0.34 (Δ)

Table 2. Characteristic nutrient levels (µM) before and after the first bloom in Arctic and Antarctic waters. (Δ) denotes the ratio for the decrease in nutrients during a bloom. Data from Ellertsen et al. (1981) and Biggs et al. (1982)

instance in midsummer from 58° at 55° lat. to 23° at the poles. Coincidentally diurnal rhythms become attenuated. This is clearly reflected in the diurnal variation in global radiation on a clear summer day (Fig. 3). Daily integrals will not differ much for clear days, when comparing different places in summertime. Weather patterns do, however, influence the global radiation considerably. Close to the Antarctic continent where high pressures prevail, the weather is often lightly cloudy or sunny, whereas the West Wind Drift is characterized by the continuous passage of low pressures and hence, dominance of cloudy weather. Thus the global radiation may be higher close to the continent than in the West Wind Drift (Holm-Hansen et al. 1977). In the Arctic summer fog and cloudy weather often prevails, but the variability is high. The monthly average global radiation is hardly more than 20-40% of that of clear days in that area (Spinnangr 1968).

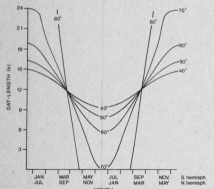

Fig. 2. Seasonal variation in day length at various latitudes. Data from Holm-Hansen et al. (1977) and Spinnangr (1968)

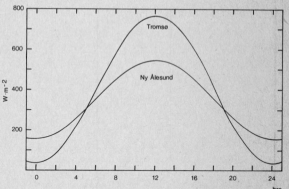

Fig. 3. Diurnal variation in global radiation on clear midsummer days at Tromsø (70°N) and Ny-Ålesund, Svalbard (79°N)

Submarine light measurements also reflect extreme seasonal variation at high latitudes. In the Trondheimsfjord, Norway (63°N), measurements with a spherical collector at 0.5m depth indicated that the peak photon flux density (aver. for the two brightest hours of the day) ranged from 30 µE m^{-2}s^{-1} in December to 1500 µE m^{-2}s^{-1} in June (Hegseth and Sakshaug 1983).

Except for situations with dense phytoplankton blooms and local turbidity from melting glacial ice, polar waters are blue and rather transparent. Hart (1962) and ACDA cruise data (Holm-Hansen and Chapman 1983) indicate a maximum Secchi depth of about 40m in Antarctic waters, which corresponds well with a maximum estimate of about 100m for the 1% photon flux density (PFD) in that area (Slawyk 1979). In the Arctic and adjacent seas the estimate for the maximum 1% PFD level is 50-60m; during blooms this decreases to 20-30m (Ellertsen et al. 1981, Sakshaug et al. 1981, Platt et al. 1982).

SPATIAL AND SEASONAL DISTRIBUTION OF POLAR PHYTOPLANKTON

In spite of half a century of studies in the Antarctic, the spatial and temporal distribution of phytoplankton is not well documented. There are several reasons for this. First, distances are vast, the seas are often rough, and expeditions are relatively few and very expensive. Second, the most extensive studies were done by the early researchers (Hart 1934, 1942, Hentschel 1932), but they generally examined only net tow samples and so included only the relatively large cells. Third, the numerous cruises of the past 20 years have been concentrated in a few geographical areas and have generally taken place during the austral summer period (see review by El-Sayed 1984). Some of the more valuable quantitative data on species distributions have been obtained by the inverted microscope method (e.g. Hasle 1969), but these studies have been relatively few and restricted in scope. The following generalizations emerge from all these studies. (1) There is a richness of phytoplankton around South Georgia, with both a spring and an autumn bloom, and the Scotia Sea also appears richer than other areas. There is, however, a delay of the spring bloom with increasing latitude due to light and ice effects. The "Northern Region" (Polar Front to 330 n.m. south of it) with generally open waters exhibits a peak in early December; the "Intermediate Region" and the Scotia Sea in January; and the "Southern Region" (south of the Antarctic Circle) exhibits a peak in mid-late February. (2) The main picture from December and onwards will consequently be one of poverty of phytoplankton in all offshore waters. Chlorophyll is generally in the range of 0.1 to 0.3 µg per liter (e.g. El-Sayed 1970b). (3) Phytoplankton blooms (>1.5 µg chl. per liter) are characteristic of shelf areas (300 to 500m depth) in the Ross and Weddell Seas, around the Antarctic Peninsula, in waters over the Scotia Ridge, and at times along receding ice egdes. Blooms in these waters commonly are over 5 µg chl. per liter.

Considering the high nutrient levels in Antarctic waters, the peak density of phytoplankton blooms do not appear impressive (Table 3) as they range far below the expected level at nutrient depletion. In sheltered waters, such as Gerlache Strait, 18-25 µg chl. l^{-1} has been observed (El-Sayed 1968, Burkholder and Mandelli 1965). This level is not much higher than that in the Arctic Ocean and adjacent seas.

Table 3. Typical ranges for chlorophyll a and primary production in Arctic and Antarctic waters which are open or seasonally ice-covered. Based on data from El-Sayed (1970b), Holm-Hansen et al. (1977) and Nemoto and Harrison (1981)

	Antarctic		Arctic	
	deep	inshore/shelf	deep	inshore/shelf
chlorophyll maximum, µg l^{-1}	8-10	18-25	10-15	10-15
expected at nutrient depletion		40-80		15-20
primary prod., g C $m^{-2} yr^{-1}$	16	25-130	25-55	50-250
primary prod., g C $m^{-2} day^{-1}$		0.05-4.7		0.2-2.5

Primary production in the Antarctic (Table 3) generally reflects the variation in phytoplankton biomass (Holm-Hansen et al. 1977), implying that the variation in growth rate is small in comparison to biomass variation. Both the daily and the annual primary production vary within wide limits with the higher values pertaining to inshore and shelf areas (Table 3). On the average higher values have been reported for the Arctic than for the Antarctic (Table 3). The seasonality of polar phytoplankton is, however, inadequately known, so it may suffice to say that the Arctic and the Antarctic are not notably different with respect to annual primary production per unit area of open or seasonally open water.

In spite of fewer investigations a general pattern for phytoplankton distribution can also be generated for the Arctic Ocean and its adjacent seas. The adjacent Norwegian Sea and the southern Barents Sea where surface water of Atlantic origin prevails, and which are permanently ice free, develop a spring bloom in early to late May (Halldal 1953, Paasche 1969, Rey 1981, Sakshaug et al. 1981) when thermal stratification sets in. Along the Norwegian Coastal Current and the fjords where stability is salinity dependent, the spring bloom occurs as early as March-April. This difference in spite of identical light regimes made Braarud and Klem (1931) postulate the importance of water column stability for spring blooms, an idea which was brought forth further by studies in the Denmark Strait and in the Greenland Sea (Braarud 1935, Steeman Nielsen 1935). On this basis Sverdrup (1953) introduced a simple mathematical model for the onset of spring blooms as well as the concept of critical depth. These ideas have been verified later by a comprehensive set of data for the Norwegian coast (Rey 1981). In these areas autumn maxima are reported for near-shore areas and fjords only. After culmination of the spring bloom nutrient depletion sets in and oligotrophy prevails till the next growth reason.

A pattern has emerged also for Arctic regions with seasonal ice cover. Epontic algae (under ice) may start growing modestly as early as February and represent a primary production of 0.015-0.020 g C m^{-2} day^{-1} (Bering Sea McRoy and Goering 1974). Later in the year, when snow on the ice melts, increased light penetration supports some growth of phytoplankton in waters under the ice. This limited growth appears unable to exhaust the nutrients of the surface waters. When the ice melts, nutrient rich surface waters thus become exposed and a vigorous bloom follows, apparently following in the wake of the retreating ice egde. This phenomenon has been reported by several authors in the Bering Sea (Alexander 1980, Schandelmeier and Alexander 1981), in the Beaufort Sea (Horner and Schrader 1982), in Frobisher Bay (Grainger 1975) and in the Barents Sea (Rey and Loeng 1984). McRoy and Goering (1974) have estimated that this bloom may contribute more than 50% of the annual primary production in the Bering Sea. A schematic illustration is given in Fig. 4. This "ice edge" bloom culminates when nutrients become depleted, with a maximum chlorophyll level in the range og 8-10 µg chl. l^{-1}. After culmination a nutrient-limited period usually sets in, of which the duration is determined by the length of the ice-free season. It may be short or non-existent at the very highest latitudes.

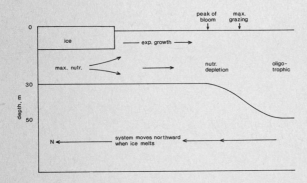

Fig. 4. Schematical illustration of the phytoplankton bloom in northern ice-edge waters. Based on data from Ellertsen et al. (1981) and Schandelmeier and Alexander (1981).

It seems rather well documented for Arctic areas that ice melting has two positive effects on the primary production: exposure of nutrient rich waters to a full strength light regime and formation of a stable water column by introduction of meltwater. Typically a pyknocline forms at 20-30m depth at the ice edge, which is about the depth of the pyknocline in Norwegian fjords during the spring bloom. In the Norwegian Sea blooms form when the pyknocline is not deeper than 40-50m. In all these cases Sverdrup's concept of critical depth may serve as an Occam's razor, i.e. explaining a maximum of phenomena with the simplest possible theory.

The scheme treated above is simplistic and real only in the most general of terms. The hydrography along the ice edge may be extremely complicated with upwelling and downwelling depending on the weather (Buckley et al. 1979, Johannesen et al. 1983), and the ice edge itself may be far from well defined for the very same reasons (Vinje 1977, Wadhams 1981). Blooms may be entirely absent if strong winds erode the stability as soon as it is formed.

The need for water column stability has also been brought into the discussion of blooms in the Antarctic Ocean. The idea was brought up first by Gran (1931) in connection with studies of the Weddell Sea where he assumed that stability was formed for a limited period by melting ice. Hart (1934, 1942) and Hasle (1969) have brought strong cases for this idea. There seems *a priori* no reason to believe that the Arctic and Antarctic Oceans are different with this respect save that in the Arctic the large amount of freshwater from rivers may also cause some additional stability. Both Hart and Hasle have explained the low phytoplankton stock at the Polar Front as a result of deep-reaching turbulence, and the delay of the spring bloom with increasing latitude may be explained by delayed melting of ice. On the VULCAN cruises (Holm-Hansen and Foster 1981) it was found that for all blooms (10 stations with >2 µg chl. l^{-1}) the pyknoline was situated at 20-40m, and 50m may be the maximum depth allowing blooms, just as in the Arctic Ocean. The importance of pyknocline depth is difficult to ascertain, however, without knowing the rate of vertical movement of the phytoplankton through the submarine light gradient (Falkowski and Wirick 1981). Unfortunately there are no convenient methods to determine the rates of such vertical motion.

When a spring bloom ends in the Arctic, it is probably due to nutrient depletion and grazing. Nutrient depletion is hardly likely in the Antarctic. Stability erosion combined with heavy grazing is a more likely proposal. It is an open question as to how long stability formed by ice melting may last. Quite obviously it is broken down in offshore waters well before a bloom approaches nutrient depletion. Depending upon the growth rate (0.4 or 0.2 div.day^{-1}) it will take a bloom 16 or 30 days to develop from 0.1 to 8 µg chl. l^{-1}. Assuming optimum conditions for rapid growth the shorter period may not be too far from reality. Based on a numerical plankton model for the Barents Sea (Slagstad 1982), it appears that sudden erosion of stability (due to a sudden storm, for instance) combined with heavy grazing and rapid sinking may break down a bloom in a matter of few days, whereas a bloom will persist for several days despite high grazing pressure and sinking rates if the breakdown of stability is gradual. Which mode of breakdown is typical for Antarctic waters is not known.

Pelagic blooms in polar oceans are usually dominated by fairly large centric diatoms (or chains of them) or colonies of the haptophyte *Phaeocystis pouchetii*. VULCAN and ACDA data have revealed that such forms make up a higher percentage of the biomass the higher the total phytoplankton biomass is. For non-bloom conditions, which are very common in the Antarctic, nanoplankton (<20nm) may constitute more than 50% of the phytoplankton biomass (Bröckel 1981). By using epifluorescence microscopy immediately after sampling (Hewes and Holm-Hansen 1983) during the ACDA cruise it was observed that about 1/3rd of the nanoplankton biomass consisted of heterotrophic organisms (Hewes et al. 1984). Thus a regenerative "heterotrophic" loop (Azam et al. 1983) may be of major importance in the Antarctic food web.

PHYSIOLOGICAL ECOLOGY OF POLAR PHYTOPLANKTON

This section deals with effects of the light regime, nutrients, and temperature on the growth rate and the chemical composition of phytoplankton. The dominant factor is the light regime, which includes the intensity, duration, and the submarine light gradient to which cells are exposed by turbulent water movements and absence/presence of ice. Physiological adaptation to varying light intensity is of considerable importance in determining growth rates of phytoplankton (Neori et al. 1984). The importance of such adaptation has also been indicated by Slagstad (1982), who found by running a numerical plankton model that shade adaptation might allow blooms in open boreal waters to set in three weeks earlier as compared to light adapted populations.

Light/shade adaptation is manifested in several ways. One is fluctuation in the cellular chlorophyll content as compared to carbon or nitrogen (Falkowski and Owens 1980). During the ACDA cruise natural water samples were incubated at ambient water temperature (0 to 4°C) and attenuated sunlight. When the PFD ($\mu E\ m^{-2} s^{-1}$) was decreased from 1050 to 6.5 (6 different light levels), the chl/N ratios increased 2.5-fold (from 0.031 to 0.072). The growth rates in these 6 cultures varied from 0.23 to 0.41 doubl. per day. The chlorophyll decrease became particularly pronounced when the light flux was inhibitory. The ATP/C ratios remained fairly constant (about 0.0043). Therefore the chl/ATP ratios also reflect light/shade adaptation. The chlorophyll levels of these cultures were quite low, however, when compared to

Skeletonema which may reach chl/N values as high as 0.25 when shade adapted (Sakshaug and Andresen, in prep.). They are also low compared to a set of cultures from the VULCAN cruise (Holm-Hansen and Foster 1981) which had chl/N ratios close to 0.1. This may reveal differences in ecological strategy for different groups of species. The ACDA cultures consisted of pennate diatoms from stable, ice-filled waters, presumably they might have been ice algae from melted ice, whereas the VULCAN 7 cultures were "monads and flagellates" and centric diatoms (*Thalassiosira* spp. and *Chaetoceros tortissimus*) from rather turbulent offshore waters.

Fluctuations in cellular chlorophyll have also been observed for natural populations. Table 4 presents data for various high latitude areas. As expected the cellular chlorophyll level is lower in surface waters than below the pyknocline (quite often by a factor of 2 or more), but we also se that cellular chlorophyll levels in the homogeneous layer tend to increase with the depth of this layer (VULCAN data). There is also a pronounced seasonal trend, quite spectacular for surface *Skeletonema*, but apparent also from the other data. With both depth and season playing a role the annual range of variation for the cellular chlorophyll level can become quite high. Hegseth and Sakshaug (1983) found that the chl/N ratio ranged through the year from 0.03 to 0.21 for *Skeletonema* and from 0.06 to 0.31 for *Thalassiosira gravida* when grown in *in situ* dialysis cultures in the Trondheimsfjord at 0.5 and 4m depth.

Light/shade adaptation may also be revealed by the shape of the P *vs*. I curve derived from short-term experiments. This has been demonstrated for Arctic as well as Antarctic phytoplankton (Burkholder and Mandelli 1965, Platt et al. 1982, El-Sayed 1984). Typically phytoplanton will exhibit different curves for different depths when waters are stratified and no difference when waters are homogeneous. Data of Platt et al. (1982) for Baffin Bay are most illustrative (Fig. 5): carbon uptake normalized to chlorophyll (assimilation number) implies a high maximum number for light adapted populations and also a somewhat higher initial slope. In terms of carbon turnover (right diagram) the picture becomes altered since the shade-adapted populations had 1.7 times more chlorophyll than the light-adapted ones, and here the superior efficiency of shade adapted populations at low light becomes evident On the other hand, both curves also reveal the high susceptibility to photoinhibition for shade-adapted populations.

Since phytoplankton is constantly exposed to light regime fluctuations, diurnal or by vertical movement, the time course of adaptation

Table 4. Chlorophyll per unit biomass for natural populations at various seasons and depth intervals in high latitudes. Hydrographical data indicate a homogeneous column within the stated depth intervals. Nitrogen and carbon has been corrected for detrital interference according to Sakshaug (1978) and Olsen et al. (1982). Data from VULCAN 6 and 7, 1981, the Trondheimsfjord (Sakshaug 1978) and the Norwegian Coastal Current (Sakshaug et al. 1981)

		m	chl/C average	no. obs.	range
Blooms, >1.5 µg chl⁻¹					
Trondheimsfjord (*Skeletonema*)	early April	0-5	0.031	5	0.029-0.036
	early May	0-5	0.025	6	0.020-0.032
	late May	0-5	0.015	9	0.013-0.018
	late May	9-10	0.044	4	0.034-0.050
Norwegian Coastal Current	late May	0-30	0.015	3	0.013-0.019
Vulcan 6 and 7	Jan-early Febr	0-40	0.014	5	0.010-0.018
	Jan-early Febr	50-75	0.025	3	0.014-0.032
Oligotrophic, <1.5 µg chl⁻¹					
Norwegian Coastal Current	late May	0-30	0.010	11	0.004-0.011
	mid May	0-30	0.015	9	0.006-0.029
VULCAN 6 and 7	Jan-early Febr	0-40	0.0060	6	0.006-0.007
	Jan-early Febr	0-50	0.0082	6	0.003-0.010
	Jan-early Febr	0-75	0.013	1	0.013
	March	0-50	0.020	7	0.014-0.029
	Jan-early Febr	50-75	0.014	5	0.007-0.020

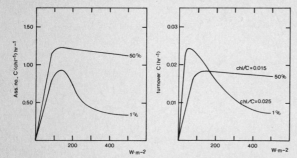

Fig. 5. Schematical P *vs*. I curves for surface phytoplankton (50% light depth) and phytoplankton below the pyknocline (1% light depth) in summer in Baffin Bay. Curves are normalized to chlorophyll (left) and carbon (right). Based on data from Platt et al. (1982).

is of major interest when studying light-governed growth. Present knowledge is sparse. Some studies suggest that temperature may be of some importance (Falkowski 1980). Platt et al. (1982) found, however, that surface populations brought to low light might take weeks to adapt in contrast to a matter of a day or two when deep populations were brought the other way. The time course seems to be related in some fashion to the expected generation time in the new regime for cells adapted to the old one. With that idea in mind the right diagram in Fig. 5 may help explain the time courses observed by Platt and colleagues and may also explain the very rapid onset of blooms when seeding stocks under the ice are exposed to full light at ice melting (Alexander, pers.comm.).

One would *a priori* expect nutrient supply to have a larger impact on phytoplankton growth in the Arctic than in the Antarctic. The question of N *vs*. P as a limiting factor is actually dependent on the species in question (Sakshaug et al. 1983). One may assume that the supply of N and P is fairly balanced relative to the needs of an "average" community since the Δ values in Table 2 are not too far from the Redfield ratio of 16. One may note that the values in Table 2 actually tend to be somewhat higher than 16, i.e. pointing towards P as a potential minimum factor. Data in Table 2 show that silica consentrations in the Arctic are very low as compared to the Antarctic. Paasche (1980) found a Si/C ratio (atoms) of 0.047 to 0.17 for 4 common species along the North Atlantic coasts, and the variation was highly species and temperature dependent. A *Skeletonema* bloom in the Oslofjord exhibited a range of 0.077 to 0.18 (Paasche and Østergren 1980). In terms of N/Si ratios the total range becomes 3.7 to 1 for non-starved cells (at N/C = 0.15, atoms) compared to 2.8 - 2 for the Δ values in Table 2. This makes it likely that some diatoms in arctic waters may be growth-limited by insufficient silica.

In the Antarctic Ocean nutrient depletion does not occur in offshore waters, and shipboard cultures may grow exponentially without enrichment until surpassing 40 µg chl liter^{-1}. There have been no chemical indications suggesting nutrient limitation in Antarctic waters. Natural populations from the Scotia Sea (VULCAN 6 and 7) and around the Antarctic continent (ACDA cruise) yielded a range for N/C of 0.11 - 0.17 with 0.15 as an average (225 samples with PON >10 µg l^{-1}), corrected for detritus according to Sakshaug (1978) and Olsen et al. (1983). N-deficiency is not evident until N/C <0.1 (Sakshaug et al. 1983). On the VULCAN 7 cruise the protein/carbohydrate ratio was measured and was always well above 5, which indicates limitation by light rather than by N and P. Nutrient deficiency yields values below unity (Myklestad 1977).

The N/Si ratio in Antarctic samples (ACDA and VULCAN cruises) have ranged between 0.5 to 4.8. It is difficult to attach much significance to these ratios as many phytoplankton groups (flagellates, e.g.) do not require silica for growth. In those samples which were dominated by diatoms, the N/Si ratio ranged from 0.48 to 1.1, values which are higher than the corresponding nutrient ratio. Data on Si-kinetics also support the view that diatom growth rates are not limited by silica availability. Jacques (1983) has reported half saturation constants of 12 µM for *Nitzschia turgidula* and 12-22 µM for the extremely silicified *Nitzschia kerguelensis*. These are very high values although nevertheless indicating growth close to the maximum rate for Si >50 µM.

In true Antarctic offshore waters the minimum level is seldom below 50 µM, indicating that Si limitation may be neglected. Near the Polar Front Si levels are considerably lower (Jacques 1983), implying possible Si-deficiency for some diatoms. Also in sheltered, Antarctic areas Si-deficiency may be possible when extremely dense blooms occur.

Recent studies (Olsen 1980, Glibert et al. 1982, Rönner et al. 1983) in the Antarctic have shown (1) that 50-80% of total nitrogen assimilated by phytoplankton is in the form of ammonia, (2) that waters with low standing stock (<1.5 µg chl liter^{-1}) have a much greater relative uptake rate of ammonia as compared to nitrate than that found in bloom conditions, and (3) that microplankton have a higher nitrate uptake relative to total N uptake than do nanoplankton. Regenerative communities (N uptake based on NH_3) are of such ubiquitous global distribution that they may be regarded as a "global plankton baseline" with blooms to be regarded as anomalies superimposed and allowed by extreme seasonality or hydrographical pecularities. The Antarctic regenerative and oligotrophic community (i.e. low standing stock) differs from others, however, in probably being the only one existing in a vast sea of nitrate, and it is difficult to explain this in terms other than energy economy in a light limited regime due to deep-reaching turbulence (50-100m). It also follows that algae belonging to that community should have an impressive ability of shade adaptation.

Antarctic phytoplankton blooms are then, in contrast, characterized by species which base their nitrogen uptake on nitrate as a main source and which have a rather high light requirement (turbulence not deeper than 20-40m). One may also speculate to which extent these species have a high energy requirement considering their high silica content. In contrast, the small *Fragilariopsis nana*-like diatoms predominant in oligotrophic communities do not appear very silicified in the microscope. Future autecological studies of characteristic species from each community may shed some light upon the mechanisms involved and possibly resolve the apparent paradox of oligotrophic communities in nutrient-rich waters.

Our photobiology studies on the ACDA and VULCAN cruises have indicated that the saturating light intensity for photosynthesis is about 10% of incident radiation, and that growth rates of 0.2 doublings per day are still obtained at intensities as low as 0.5% for I_o. Growth rates of Antarctic phytoplankton under optimum light conditions range from 0.1 to 0.9 doublings per day (Jacques 1983, Bunt and Lee 1970, Holm-Hansen et al. 1977, Jacques and Minas 1981, El-Sayed and Taguchi 1981, VULCAN and ACDA data). These growth rates are generally well below the maximum rates predicted by Eppley's (1972) equation (0.75 to

1.1 doubl. day for -1 to 4°C). They are also generally below the rates predicted by equations based on growth of *Skeletonema* in outdoor dialysis cultures (~ 80% of Eppley's rates, Sakshaug 1977, Hegseth and Sakshaug 1983). Their cold adaptation consists of being obligate psychrophilic (Neori and Holm-Hansen 1982, Jacques 1983) and apparently not in being able to grow fast at low temperatures (Fogg 1977). In this respect they differ strikingly from some species which form blooms in the Arctic. *Thalassiosira nordenskioeldii* and *Detonula confervacea*, although not being strictly Arctic, belong to this category. They grow at their optimum temperature at least as fast as predicted by Eppley's equation, and they are not as strongly obligate psychrophilic as Antarctic diatoms investigated hitherto (Smayda 1969, Durbin 1974, Neori and Holm-Hansen 1983). It thus seems that the rate of photosynthesis per unit phytoplankton biomass in Antarctic waters is severely limited by thermo-dynamic effects on cellular enzyme systems.

REFERENCES

Alexander V (1980) Interrelationships between the seasonal sea ice and biological regimes. Cold Regions Sci. Technol. 2 : 157-178.

Azam F, Field JG, Gray JS, Meyer-Reil LA and Thingstad F (1983) The ecological role of water-column microbes in the sea. Mar. Ecol. Progr. Ser. 10 : 257-263.

Biggs DC, Johnson MA, Bidigare RR, Guffy JD and Holm-Hansen O (1982) Shipboard autoanalyzer studies of nutrient chemistry, 0-200m, in the Eastern Scotia Sea during FIBEX (January-March 1981). Techn Rept 82-11-T, Dept. Oceanography, Texas A & M University, p 98.

Braarud T (1935) The "Øst" expedition to the Denmark Strait 1929. II. The phytoplankton and its conditions of growth. Hvalråd. Skr. 10 : 1-173.

Braarud T and Klem A (1931) Hydrographical and chemical investigations in the coastal waters off Møre and in the Romsdalsfjord. Hvalråd, Skr. 1 : 1-88.

Bröckel K von (1981) The importance of nanoplankton within the pelagic Antarctic ecosystem. Kieler Meeresforsch. Sonderheft 5 : 61-67.

Buckley JR, Gammelsrød T, Johannesen JA, Johannesen OM and Røed LP (1979) Upwelling: Oceanic structure at the edge of the Arctic ice pack in winter. Science, N.Y. 203 : 165-167.

Bunt JS and Lee CC (1970) Seasonal primary production in antarctic sea ice at McMurdo Sound in 1967. J. Mar. Res. 28 : 304-320.

Burkholder PR and Mandelli EF (1965) Carbon assimilation of marine phytoplankton in Antarctica. Proc. National Acad. Sci. (US) 54 : 437-444.

Durbin EG (1974) Studies on the autecology of the marine diatom *Thalassiosira nordenskioeldii* Cleve. 1. The influence of day-length, light intensity and temperature on growth. J. Phycol. 10 : 220-225.

Ellertsen B, Loeng H, Rey F and Tjelmeland S (1981) The feeding condition of capelin during summer. Field observations in 1979 and 1980. Fisken Hav. 1981 (3) : 1-68 (in Norwegian, English abstract).

El-Sayed SZ (1968) On the productivity of the southwest Atlantic Ocean and the waters west of the Antarctic Peninsula. In: Llano AG and Schmitt WK (eds.) Biology of the Antarctic Seas III. Antarct. Res. Ser. 11 : 15-47.

El-Sayed SZ (1970a) Biological aspects of the pack ice ecosystem. Proc. Symp. Antarct. Ice and Water Masses, Tokyo : 35-54.

El-Sayed SZ (1970b) On the productivity of the Southern Ocean. In: Holdgate MW (ed.) Antarctic Ecology 1 : 119-135. Academic Press, London and New York.

El-Sayed SZ (1984) Biological productivity of the Antarctic waters - present paradoxes and paradigms. Proc. Symp. Reg. Aq. Biol. Antarctica, San Carlos de Bariloche Argentina, June 83 (in press).

El-Sayed SZ and Taguchi S (1981) Primary production and standing crop of phytoplankton along the ice-edge in the Weddell Sea. Deep-Sea Res. 28 : 1017-1032.

Eppley RW (1972) Temperature and phytoplankton growth in the sea. Fishery Bull. NOAA 70 : 1063-1085.

Falkowski PG (1980) Light-shade adaptation in marine phytoplankton. In: Falkowski PG (ed.) Primary productivity in the sea : 99-119. Plenum Press, London and New York.

Falkowski PG and Owens TG (1980) Light-shade adaptation. Plant Physiol. 66 : 592-595.

Falkowski PG and Wirick CD (1981) A simulation model of the effects of vertical mixing on primary productivity. Mar. Biol. 65 : 69-75.

Fogg GE (1977) Aquatic primary production in the Antarctic. Phil. Trans. R. Soc. Lond. B 279 : 27-38.

Glibert PM, Biggs DC and McCarthy JJ (1982) Utilization of ammonium and nitrate during austral summer in the Scotia Bay. Deep-Sea Res. 29 : 837-850.

Grainger EH (1975) A marine ecology study in Frobisher Bay, Arctica Canada. In: Billingsby LW and Cameron TWM (eds.) Energy flow - its biological dimensions. A summary of the IBP in Canada, 1964-1974. Canadian Committee for the IBP, Roy Soc. Can.

Gran HH (1931) On the conditions for the production of plankton in the sea. Cons. perm. int. Explor. Mer 75 : 37-46.

Gulland JA (1970) The development of resources of the Antarctic seas. In: Holdgate MW (ed.) Antarctic ecology, 1 : 217-224. Academic Press, London and New York.

Halldal P (1953) Phytoplankton investigations from weather ship M in the Norwegian Sea, 1948-49. Hvalråd. Skr. 38 : 1-91.

Hart TJ (1934) On the phytoplankton of the southwest Atlantic and the Bellingshausen Sea, 1929-1931. Discovery Repts 8 : 1-268.

Hart TJ (1942) Phytoplankton periodicity in antarctic surface waters. Discovery Repts 21 : 261-365.

Hart TJ (1962) Notes on the relation between transparency and plankton content of the surface waters of the Southern ocean. Deep-Sea Res. 9 : 109-114.

Hasle GR (1969) An analysis of the phytoplankton of the Pacific Southern Ocean: abundance, composition, and distribution during the Brategg expedition, 1947-1948. Hvalråd Skr. 52 : 1-168.

Hegseth EN and Sakshaug E (1983) Seasonal variation in light- and temperature-dependent growth of marine plankton diatoms in $in\ situ$ dialysis cultures in the Trondheimsfjord, Norway (63^ON). J. exp. mar. Biol. Ecol. 67 : 199-220.

Hentschel E (1932) Deutsche Atlantische Expedition, Meteor 1925-27. Vol. 10, 274 pp, W. de Gruyter and Co., Berlin.

Hewes C and Holm-Hansen O (1983) A method for recovering nanoplankton from filters for identification with the microscope: The filter-transfer-freeze (FTF) technique. Limnol. Oceanogr. 28 : 389-394.

Hewes CD, Holm-Hansen O and Sakshaug E (1984) Alternate carbon pathways at lower-trophic levels in the antarctic food web. Proc. 4th Symp. Antarct. Biol., Wilderness, South-Africa (in press).

Holm-Hansen O, El-Sayed SZ, Franceschini GA and Cuhel RL (1977) Primary production and the factors controlling phytoplankton growth in the Southern Ocean. In: Llano GA (ed.) Adaptation within antarctic ecosystems, Proc. 3d SCAR Symp. Antarct, Biol. : 11-50.

Holm-Hansen O and Foster TD (1981) A multidisciplinary study of the eastern Scotia Sea. Antarctic J. US 16 : 159-160.

Holm-Hansen O and Huntley M (1984) Feeding requirements of krill in relation to food sources. J. Crustacean Biology (in press).

Holm-Hansen O and Chapman AS (1983) Antarctic circumnavigation cruise, 1983. Antarct. J. US (in press).

Horner R (1976) Sea ice organisms. Oceanogr. Mar. Biol. Ann. Rev. 14 : 167-182.

Horner R (1977) History and recent advances in the study of the ice biota. In: Dunbar MJ (ed.) Polar Oceans, 269-283. Arctic Inst. North America.

Horner R and Schrader GS (1982) Relative contributions of ice algae, phytoplankton, and benthic microalgae to primary production in nearshore regions of the Beaufort Sea. Arctic 35 : 485-503.

Jacka TH (1981) Antarctic temperature and sea ice extent studies. In: Young NW (ed.) Antarctica : weather and climate. Preprint volume, Univ. Melbourne, 10 p.

Jacques G and Minas M (1981) Production primaire dans le secteur indien de l'Ocean Antarctique en fin d'été. Oceanol. Acta 4 : 33-41.

Jacques G (1983) Some ecophysiological aspects of the Antarctic phytoplankton. Polar Biol. 2 : 27-33.

Johannesen OM, Johannesen JA, Morison J, Farelly BA and Svendsen EAS (1983) Oceanographic conditions in the marginal ice zone north of Svalbard in early fall 1979 with an emphasis on mesoscale processes. J. geophys. Res. 88 (C5) : 2755-2769.

McRoy CP and Goering JT (1974) The influence of ice on the primary productivity of the Bering Sea. In: Hood D and Kelley E (eds.) The Oceanography of the Bering Sea : 403-421. Univ. Alakska Inst. mar. Sci.

Myklestad S (1977) Production of carbohydrates by marine planktonic diatoms. II. Influence of the N/P ratio in the growth medium on the assimilation ratio, growth rate, and production of cellular and extracellular carbohydrates by *Chaetoceros affinis* var. *willei* (Gran)Hustedt and *Skeletonema costatum* (Grev.)Cleve. J. exp. mar. Biol. Ecol. 29 : 161-179.

Nemoto T and Harrison G (1981) High latitude ecosystems. In: Longhurst AR (ed.) Analysis of marine ecosystems : 95-126. Academic Press, London and New York.

Neori A and Holm-Hansen O (1982) Effect of temperature on rate of photosynthesis in Antarctic phytoplankton. Polar Biology 1 : 33-38.

Neori A, Holm-Hansen O, Mitchell BG and Kiefer DA (1984) Photoadaptation in marine phytoplankton: changes in spectral absorption and excitation of chlorophyll a fluorescence. Plant Physiology (submitted).

Olsen Y, Jensen A, Reinertsen H and Rugstad B (1983) Comparison of different algal carbon estimates by use of the Droop-model for nutrient limited growth. J. Plankton Res. 5 : 43-51.

Olson RJ (1980) Nitrate and ammonium uptake in Antarctic waters. Limnol. Oceanogr. 25 : 1064-1074.

Paasche E (1960) Phytoplankton distribution in the Norwegian Sea in June, 1954, related to hydrography and compared with primary production data. FiskDir. Skr. Ser. Havunders. 12 : 1-77.

Paasche E (1980) Silicon content of five marine plankton diatom species measured with a rapid filter method. Limnol. Oceanogr. 25 : 474-480.

Paasche E and Østergren I (1980) The annual cycle of plankton diatom growth and silica production in the inner Oslofjord. Limnol. Oceanogr. 25 : 481-494.

Platt T, Harrison WG, Irwin B, Horne EP and Gallegos CL (1982) Photosynthesis and photoadaptation of marine phytoplankton in the Arctic. Deep-Sea Res. 29 : 1159-1170.

Rey F (1981) The development of the spring phytoplankton outburst at selected sites off the Norwegian coast. In: Sætre R and Mork M (eds.) The Norwegian coastal current, 649-680. University of Bergen.

Rey F and Loeng H (1984) The influence of ice and hydrographic conditions on the development of phytoplankton in the Barents Sea. Proc. 18th Eur. Symp. Mar. Biol. Oslo 1983 (in press).

Rönner U, Sörensen F and Holm-Hansen O (1984) Nitrogen assimilation by phytoplankton in the Scotia Sea. Polar Biol. (in press).

Sakshaug E (1977) Limiting nutrients and maximum growth rates for diatoms in Narragansett Bay. J. exp. mar. Biol. Ecol. 28 : 109-123.

Sakshaug E (1978) The influence of environmental factors of the chemical composition of cultivated and natural populations of marine phytoplankton. D.Sc.thesis, University of Trondheim. 82 p.

Sakshaug E, Myklestad S, Andresen K, Hegseth EN and Jørgensen L (1981) Phytoplankton off the Møre Coast in 1975-1976: distribution, species composition, chemical composition and conditions for growth. In: Sætre R and Mork M (eds.) The Norwegian coastal current : 681-711. University of Bergen.

Sakshaug E, Andresen K, Myklestad S and Olsen Y (1983) Nutrient status of phytoplankton communities in Norwegian waters (marine, brackish and fresh) as revealed by their chemical composition. J. Plankton Res. 5 : 175-196.

Schandelmeier L and Alexander V (1981) An analysis of the influence of ice on spring phytoplankton population structure in the south-east Bering Sea. Limnol. Oceanogr. 26 : 935-943.

Slagstad D (1982) A model of phytoplankton growth - effects of vertical mixing and adaptation to light. Modeling, Identification and Control 3 : 111-130.

Slawyk G (1979) ^{13}C and ^{15}N uptake by phytoplankton in the Antarctic upwelling area: Results from the Antiprod I cruise in the Indian Ocean sector. Austr. J. mar. freshw. Res. 30 : 431-448.

Smayda T (1969) Experimental observations on the influence of temperature, light, and salinity on cell division of the marine diatom, *Detonula confervacea*. J. Phycol. 5 : 150-157.

Spinnangr G (1968) Global radiation and duration of sunshine in Northern Norway and Spitsbergen. Meteorol. Ann. Oslo 5 : 65-137.

Steemann-Nielsen E (1935) The production of phytoplankton at the Faroe Isles, Iceland, East Greenland and in waters around. Medd. Komm. Danm. Fisk. Havunders. Ser. Plankton 3 (1) : 1-93.

Sverdrup HU (1953) On conditions for the vernal blooming of phytoplankton. J. Cons. perm. int. Explor. Mer 18 : 287-295.

Treshnikov AF and Baranov GI (1977) The structure of the circulation and budget dynamics of the waters of the north Polar region. In: Dunbar MJ (ed.) Polar Oceans : 33-44. Arctic Inst. North America.

Vinje TE (1977) Sea ice conditions in the European sector of the marginal seas of the Arctic, 1966-75. Årbok Norw. PolarInst. 1975 : 163-174.

Wadhams P (1981) The ice cover in the Greenland and Norwegian seas. Rev. Geophys. Space Phys. 19 : 345-393.

Walsh JE and Johnson CM (1979) Analysis of arctic sea ice fluctuations. J. Phys. Oceanogr. 9 : 1953-1977.

PRODUCTIVITY OF THE ANTARCTIC WATERS - A REAPPRAISAL

S. EL-SAYED

Department of Oceanography, Texas A & M University,
College Station, Texas 77843, USA

I. INTRODUCTION

The study of Antarctic marine phytoplankton has a long history that dates back nearly a century and a half, when J. D. Hooker, the famed botanist-surgeon of the EREBUS and TERROR Expedition (1839-43) reported the ubiquitous presence of diatoms during the Antarctic summer. Hooker sent some of the samples collected between Cape Horn and the Ross Sea to C. G. Ehrenberg, who published the first paper on Antarctic diatoms in 1844.

During the following three-quarters of a century, extensive collecting (mostly using nets) of Antarctic phytoplankton was conducted by members of such celebrated expeditions as the CHALLENGER (1872-76), ANTARCTICA (1895), BELGICA (1897-99), VALDIVIA (1898-99), SOUTHERN CROSS (1899-1900), GAUSS (1901-03), SCOTIA (1902-04) and the POURQUOI PAS? (1908-10). The majority of the papers published dealt with diatoms and, to a much lesser extent, with dinoflagellates and silicoflagellates.

The initiation of the DISCOVERY Investigations (1925-39), which laid the foundation of our knowledge of the general oceanography of the Southern Ocean, ushered in a new phase of Antarctic phytoplankton research. Hart (1934, 1942) made one of the most extensive and detailed studies of Antarctic phytoplankton ever conducted. Despite his nonquantitative methods, Hart was able to describe the gross features of phytoplankton periodicity in Antarctic waters better than any other student of Antarctic phytoplankton (Hasle, 1969). Hart divided the Antarctic Zone into three different regions, and grouped the species collected according to their seasonal and geographic distribution. He observed that the main increase in phytoplankton started earlier in the Northern Region than it did further south. Using cell numbers and settled volumes and what he referred to as the

"Harvey plant pigment unit", Hart was able to estimate the standing crop of Antarctic phytoplankton.

In the early 50's, the use of the chlorophyll *a* method in estimating phytoplankton standing crop (Richards and Thompson, 1952) and the radioactive ^{14}C uptake method in estimating primary production (Steemann Nielsen, 1952) revolutionized phytoplankton research. Thanks to the extensive investigations carried out by scientists aboard the USSR OB and VITIAZ, the USNS ELTANIN, the Argentine SAN MARTIN and ISLAS ORCADAS, the French COMMANDANTE-CHARCOT and MARION DUFRESNE, the numerous Japanese Antarctic Research Expeditions, and many others in the 60's and 70's, a large amount of useful data were obtained on the geographical and temporal distributions of phytoplankton standing crop and on the magnitude of primary production in the circum-Antarctic waters (Klyashtorin, 1961; Ichimura and Fukushima, 1963; Mandelli and Burkholder, 1966; El-Sayed, 1967, 1968).

In this paper we will review the subject of primary productivity of the Antarctic waters and examine the factors which govern the productivity of these waters in light of the more recent data collected during the last decade. Special emphasis will be put on the pelagic primary producers. (See El-Sayed, 1970; El-Sayed and Turner, 1977; Fogg, 1977; Holm-Hansen *et al.*, 1977; for earlier reviews.)

II. STANDING CROP AND PRIMARY PRODUCTION OF ANTARCTIC PHYTOPLANKTON

The bulk of the phytoplankton data show chlorophyll concentrations in the range of 0.1 and 1.0 mg/m^3, with the mean value of < 0.5 mg/m^3 (Saijo and Kawashima, 1964; El-Sayed and Mandelli, 1965; El-Sayed, 1970; Fukuchi, 1980). In contrast to the generally oligotrophic oceanic regions, high standing crop values have been frequently reported from inshore waters, e.g. west of the Antarctic Peninsula, in the Gerlache Strait (El-Sayed, 1967), in the southern Ross Sea (El-Sayed *et al.*, 1983), near the Kerguelen and Heard Islands (El-Sayed and Jitts, 1973), off the Crozet Archipelago (El-Sayed *et al.*, 1979) and in the inshore waters of Signy Islands (Horne *et al.*, 1969). Exceptionally high values (i.e. in excess of 25 mg/m^3) were reported by Mandelli and Burkholder (1966), during a phytoplankton bloom near Deception Island. The most extensive and richest of these blooms, reported by El-Sayed (1971), occurred in the southwestern Weddell Sea. This bloom (composed entirely of the diatom Thalassiosira tumida (Janish) Hasle was to cover an area > 15,000 km^2, with a chlorophyll concentration of about 190 mg/m^3. Another, but less spectacular bloom (entirely dominated by the colonial flagellate Phaeocystis pouchetti (Hariot) Langerheim was widespread over the southern Ross Sea region (nearly 620 km), and extended from the Ross Ice Shelf offshore to about 185 km. At some stations, P. pouchetti reached depths of 100 to 150 m. Another large bloom of P. pouchetti was encountered during the German Antarctic Expedition (November-December, 1980) in the vicinity of the South Shetland Islands in the Bransfield Strait; the average concentration of chlorophyll was between 5-11 mg/m^3 (von Bodungen, et al., 1982).

The vertical distribution of chlorophyll a shows that maximum values were found at subsurface depths. Following these maxima, there is a gradual decrease in the chlorophyll values to the depth of 200 m, below which chlorophyll concentration is greatly reduced (El-Sayed, 1970; El-Sayed and Turner, 1977; El-Sayed and Weber, 1982). At several stations in the Pacific and Indian sectors of the Antarctic, substantial amounts of chlorophyll were found below the euphotic zone.

Primary productivity data from the Southern Ocean, in general, showed good correlation with the distribution of the phytoplankton standing crop. For instance, low values were reported in the Drake Passage, the Bellingshausen Sea, and in the oceanic waters in general. The open-ocean system recorded rates of production typical of oligotrophic regions (~ 0.1 gC/m^2/day). On the other hand, high values were found in coastal regions and in the vicinity of Antarctic and sub-Antarctic islands. For example, El-Sayed (1967) recorded 3.2 gC/m^2/day in the Gerlache Strait; Mandelli and Burkholder (1966) reported 3.62 gC/m^2/day near Deception Island. Horne et al. (1969) found a peak of productivity of 2.8 gC/m^2/day in the inshore waters of Signy Island in the South Orkney Islands. These values which are comparable to the upwelling systems off Peru, southeast Arabia, Somalia and southwest Africa no doubt have perpetuated the belief in the proverbial richness of the Antarctic waters. They may also have contributed to the high productivity estimates of the Antarctic waters published in the mid-sixties.

As to the vertical distribution of primary productivity in the circum-Antarctic waters, maximum photosynthetic activity occurred at depths corresponding to between 25 and 50% of surface light intensity. At most of the stations where primary production measurements were carried out, ^{14}C uptake extended well below the depth of 1% surface illumination (i.e. the euphotic zone). The contribution of primary production below the euphotic zone in the Ross and Weddell Seas was nearly one-fourth of the total production in the water column (El-Sayed and Taguchi, 1981; El-Sayed et al., 1983). However, in the Atlantic and Indian sectors of the Antarctic, carbon fixation below the euphotic zone did not exceed 5 to 10% of that in the entire water column (El-Sayed and Weber, 1982; El-Sayed and Jitts, 1973).

Based on the in situ primary productivity experiments conducted by the author during ELTANIN Cruises 38, 46 and 51 (Fig. 1), a mean primary productivity value of 0.134 gC/m^2/day was calculated (El-Sayed and Turner, 1977).

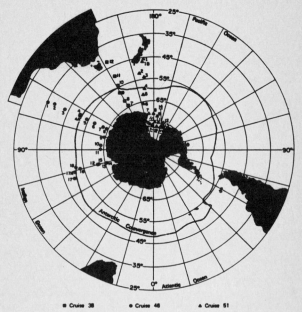

Fig. 1. Positions of stations occupied during Eltanin cruises 38 (22 March to 10 May 1968), 46 (22 November 1970 to 19 January 1971), and 51 (17 January to 24 February 1972).

The results of the phytoplankton investigations carried out in the 60's and 70's have contributed a great deal of valuable information on the standing crop of phytoplankton and primary production of the Antarctic waters. These studies were to underscore the following: (a) the productivity of these waters varies by at least one to two orders of magnitude; and (b) Antarctic waters, in general, display their richness mainly in coastal regions (El-Sayed and Turner, 1977). These studies succeeded also in dispelling the notion of the "proverbial" richness of the Southern Ocean.

Although much useful information has been gathered on the physiological ecology of phytoplankton during the past two decades, we still know very little about the metabolic activities of Antarctic phytoplankton, and whether or not they are well adapted to their environment is still a matter of controversy (Holm-Hansen et al., 1977; Olson, 1980; El-Sayed and Taguchi, 1981). Further, we still have a poor understanding of the factors which govern phytoplankton production in the Southern Ocean. Investigators are confounded by the fact that the ocean south of the Polar Front contains a circumpolar phytoplankton population living in a fairly uniform environment with sufficient light for photosynthesis (at least during spring and summer), a population that is more or less adapted to cold temperature and has abundant nutrient salts-- yet that population is not only patchy but very low in density, approaching that of oliogtrophic regions. The nutrient levels should be able to support a phytoplankton biomass of at least 25 µg chl a/l (Holm-Hansen and Huntley, in press). However, the Antarctic phytoplankton population seldom attains this density, except for the occasional blooms referred to above.

III. FACTORS GOVERNING PRIMARY PRODUCTION IN THE SOUTHERN OCEAN

El-Sayed and Mandelli (1965) discussed five factors which they considered likely to control production in the Antarctic waters. They listed: (1) light, (2) temperature, (3) inorganic nutrient salts, (4) turbulence and (5) grazing. In this section an attempt will be made to update the information given in El-Sayed and Mandelli (ibid.) in light of the data/information that has accumulated in recent years.

A. Light

Variations in incoming solar radiation between summer and winter in the Antarctic are extreme. The effects these variations have on marine plant life in the circum-Antarctic waters were discussed by El-Sayed (1971) and are illustrated in Figure 2. Furthermore, light penetration in Antarctic waters is determined not only by the intensity and angle of incidence of light, surface reflectance, and the absorption of suspended particles, but also by the presence of thick fast ice and pack-ice which appreciably reduces the amount of submarine illumination. Jacques (1983) called attention to the fact that the effect of fluctuation in the light actually received by Antarctic phytoplankton has been totally overlooked. He listed the effects of gales (time scale in hours), of passing clouds (time scale in minutes) and high frequency waves (time scale in seconds) as factors which influence the vertical movement and dispersal of phytoplankton and to which Antarctic phytoplankton can hardly adjust their physiology.

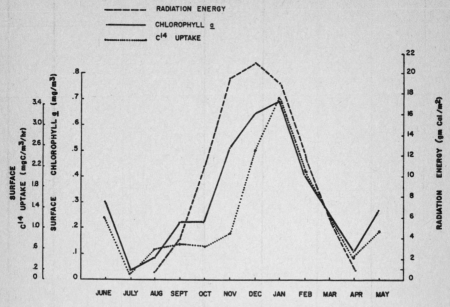

Fig. 2. MONTHLY VARIATION OF RADIATION ENERGY (WITH AVERAGE CLOUDINESS) AT "MAUDHEIM" STATION (71°03'S, 10°56'W) COMPARED WITH MONTHLY CHANGES IN SURFACE CHLOROPHYLL a AND C¹⁴ UPTAKE IN ANTARCTIC WATERS.

In their study of the effect of radiant energy on the photosynthetic activity of Antarctic phytoplankton, Holm-Hansen et al. (1977) found a high degree of correlation between the intensity of solar radiation and the photosynthetic rates in the euphotic zone. For instance, they found that on days when light intensity was high, photosynthetic rates were low in the surface waters, and these rates increased with depth. On the other hand, when incident light was low during the in situ incubation period, photosynthetic rates remained fairly constant in the upper waters of the euphotic zones or they were highest in surface waters (Fig.3). These results are most probably due to photo-inhibition, whereby high intensity light causes a decrease in the photosynthetic rates. Holm-Hansen et al. (1977) found that on bright days (100 to 160 cal/cm^2/half-light day), maximum photosynthetic assimilation occurred at depths corresponding to between 25 and 50% of incident radiation. When, on the other hand, solar radiation was low (20 to 30 cal/cm^2/half-light day), there was no evidence of photo-inhibition. The threshold of photo-inhibition for Antarctic phytoplankton was calculated to be in the range of 40 to 50 cal/cm^2/half-light day. It should be remembered that during the austral summer months, total daily flux can exceed that of tropical latitudes (Holm-Hansen et al., 1977).

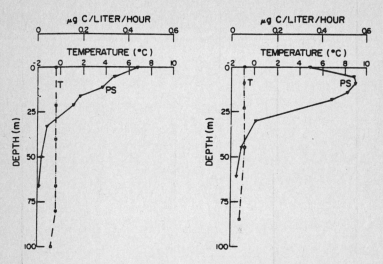

FIG. 3 VERTICAL DISTRIBUTION OF PHOTOSYNTHESIS AND TEMPERATURE AT STATION 14 (LEFT) AND STATION 16 (RIGHT) OF ELTANIN CRUISE 51 (JANUARY/FEBRUARY, 1972) IN THE ROSS SEA. PAR (FOR STATION 14): 32 CAL CM^{-2}HALF-LIGHT-DAY^{-1}
PAR (FOR STATION 16): 72 CAL CM^{-2}HALF-LIGHT-DAY^{-1}

Holm-Hansen and Sakshaug (personal communication) showed that the photo-chemical apparatus is "saturated" between 100 and 180 μ Einstein/m^2/sec., depending on the depth from which the sample was obtained. As the incident light flux (on a sunny day) is about 2,500 μ Einstein/m^2/sec., it is evident that phytoplankton in surface waters will be either "saturated" by the available flux, or will be photo-inhibited. Other investigators (Jacques, 1983) found that the metabolism of Antarctic phytoplankton does not show unusual characteristics. In his P vs I experiments, Jacques (ibid.) showed that the photosynthetic parameters (i.e. initial slope, α, and light intensity at the onset of light saturation, I_k) gave very low values for both the natural communities and algal cultures. He contends that during the summer months, "these low values of I_k express clearly the plankton's inability to use optimally the available light in the euphotic layer". Further, the very low assimilation number (AN) for Antarctic phytoplankton compared to phytoplankton from temperate and tropical upwelling regions has been attributed to the low photosynthetic efficiency of Antarctic phytoplankton. Low

I_k further indicates the influence of temperature as a controlling factor in the growth of Antarctic phytoplankton, as explained below.

B. Temperature

Temperature is commonly listed as one of the main factors, or the main factor (Saijo and Kawashima, 1964), influencing the rate of primary production in Antarctic waters. Antarctic phytoplankton show a sharp drop in photosynthetic rates at temperatures above 10°C. Thus it would seem that these algal cells are obligate psychrophiles since they grow well at low temperatures, and optimally at less than 5°C, but they will not grow at higher temperatures. Moreover, the specific growth rates (μ) for Antarctic phytoplankton are of the order of 0.1 to 0.3 doubling per day in the Ross Sea (Holm-Hansen et al., 1977) and 0.4 and 0.6 doubling per day in the southern part of the Indian Ocean (Jacques and Minas, 1981). The highest μ recorded (0.71) by El-Sayed and Taguchi (1981) from the Weddell Sea is comparable to the maximum division rate (0.75) to be expected at -2°C (Eppley, 1972). However, according to Jacques (1983) this μ still falls short of the optimum rate derived from Arrhenius law (greater than 1 doubling per day at 5°C). It further indicates that the low temperatures of Antarctic waters do indeed limit algal growth rates. El-Sayed and Taguchi (1981) data thus support the hypothesis than Antarctic phytoplankton are physiologically adapted to exhibit near maximal growth rates at low temperatures.

Recent data from the Scotia Sea/northern Weddell Sea have also indicated the importance of temperature as a controlling factor for photosynthetic rates of Antarctic phytoplankton (Neori and Holm-Hansen, 1982). These authors conclude that temperature does limit primary production rates at times when light intensity is saturating the photochemical apparatus of the cell. Since the phytoplankton are saturated by light intensity which is approximately 10-15% of that generally incident upon the sea surface, it is apparent that temperature can be a rate-controlling factor in the upper 10-20 m of the water column.

C. Nutrient Salts

The numerous observations on the nutrient salts in the Antarctic waters clearly show that these salts appear to be in excess of phytoplankton requirements. There are no data available to suggest that phytoplankton growth in the Antarctic is limited by nutrient deficiency.

Even at the peak of phytoplankton growth, the concentration of nutrient salts remains well above the limiting values. For instance, during the heavy bloom of phytoplankton in the southwestern Weddell Sea, mentioned earlier, the concentrations of nutrient salts were still high (El-Sayed, 1968). During the Phaeocystis pouchetti bloom encountered by El-Sayed et al.(1983) in the Ross Sea, nitrate concentration for instance was still high, with euphotic zone concentrations averaging 17 µg at./l. Thus it is unlikely that these nutrient salts are sufficiently low at any one time to become limiting factors to the growth of the phytoplankton. However, according to Walsh (1971), and Allanson et al. (1981), the pattern of silicate distribution indicates that SiO_3 may be the most limiting of the major nutrients for the growth of Antarctic diatoms, thus corroborating a similar conclusion arrived at earlier by Hart (1942).

With regard to the trace elements, Jacques (1983) carried out enrichment experiments using Zn, Mo, Co, Mn, and Fe, and showed that they are not limiting factors. It is possible, however, that organic factors, e.g. vitamin B_{12} and thiamin (see Carlucci and Cuhel, 1977) may alter the species composition of the phytoplankton without changing the overall rate of primary production. It is also possible that the availability of trace/micronutrients may be altered by pack-ice and iceberg melt-water, thus affecting the productivity or species composition of the waters in the vicinity of pack-ice and icebergs.

El-Sayed et al. (1964), El-Sayed and Jitts (1973) and El-Sayed and Weber (1982) reported elevated values in the standing crop and primary production in the Antarctic coastal waters and in the vicinity of Antarctic and subantarctic islands. These authors attributed this increase to the so-called "island-mass effect" (Doty and Oguri, 1956) and the attendant increase in nutrient concentrations. However, increased concentrations of inorganic salts cannot be invoked here, as is usually the case in tropical or temperate waters, since these nutrients as mentioned above, are in abundant supply. Nor are the elevated values of the phytoplankton standing crop and primary production in the Southern Ocean limited to the Antarctic coastal waters and the vicinity of islands. During ISLAS ORCADAS Cruises 17 and 19 in the southwestern Atlantic and the Scotia Sea, several deep stations were noted for their rich phytoplankton populations (El-Sayed and Weber, 1982). Further, many other deep oceanic stations in the Atlantic, Pacific and Indian sectors of the Antarctic also exhibited higher-than-average values of chlorophyll a and primary production. For example, at a station in the Weddell Sea located 600-800 km from the coast and

> 4,000 m deep, an immensely rich diatom assemblage of over 1.4×10^6 cells/l was reported (Fryxell and Hasle, 1971). These examples of Antarctic richness are not representative of the overall phytoplankton activity; they remain the exception. Nevertheless, they point out that the Southern Ocean is capable of higher levels of standing crop given its rich supply of nutrient salts.

D. <u>Water Column Stability</u>

Several investigators have drawn attention to the importance of the stability of the water column in controlling production (Braarud and Klem, 1931; Gran, 1931; Sverdrup, 1953; Pingree, 1978). Of the several processes which have been hypothesized to be important in initiating and sustaining near-ice bloom, the most important is the vertical stability induced by melt-water. According to this hypothesis, first proposed by Marshall (1957) for Arctic waters, the low salinity of melt-water contributes to the stability of the near-ice water column, thus helping to retain the phytoplankton in near surface, photic water and to promote a bloom. There is evidence that this mechanism is important in the initiation of ice-edge blooms in the Antarctic Ocean as well (Jacobs and Amos, 1967; El-Sayed, 1971). Thus stability of the upper layers plays an important role in the development of Antarctic phytoplankton blooms. Sakshaug and Holm-Hansen (personal communication) reported that for all the "bloom stations" where chlorophyll <u>a</u> was > 2 mg/m^3, the pycnocline was between 20 and 40 m deep. They speculate that 50 m is the maximum depth for the bloom to develop. They contend that the presence of homogenous (i.e. isothermal) layers, reaching to 50-100 m depth for most of the year, hinders the development of a bloom and contributes to the low primary production of the Antarctic waters.

On the other hand, the Polar Front has been cited by several investigators as a factor contributing to a low standing crop of phytoplankton. This has been explained as a result of comparatively low vertical stability of the water column, which prevents the phytoplankton from remaining in the optimal light zone long enough for extensive production (Hart, 1942; El-Sayed and Mandelli, 1965; Hasle, 1969). Paradoxically, during Cruises 17 and 19 of the ISLAS ORCADAS, higher-than-average values of chlorophyll <u>a</u> and primary production were found in the vicinity of the Polar Front. Allanson <u>et al</u>. (1981) noted marked increases in primary production and phytoplankton biomass at the Polar Front in the southwestern part of the Indian Ocean; and Yamaguchi and Shibata (1982) recorded high chlorophyll <u>a</u> at the Polar Front south of

Australia on a south-bound transect (however, they also reported low concentrations at the Polar Front during the north-bound leg of the same cruise). Interestingly, Kennett (1977) contends that the high rates of siliceous biogenic sedimentation at the Polar Front may be indicative of substantial primary production in the surface waters of this region. If the Antarctic waters were generally well mixed vertically, the instability of the water column within the Polar Front would not have as pronounced an effect on chlorophyll $\underline{a}$ concentration as has been previously suggested. Thus, the high standing crop and primary productivity at the Polar Front will have to be explained in terms other than the stability or instability or instability of the water column. At present it is difficult to speculate on the factors contributing to the high standing crop of phytoplankton observed at the Polar Front region.

E. Grazing

In the Southern Ocean, euphausiids (principally Euphausia superba), which may constitute half of the Antarctic zooplankton (Holdgate, 1970), are the dominant herbivores. Several investigators have observed that areas of high krill concentration are usually noted for their low standing crop of phytoplankton. During the First International BIOMASS Experiment (FIBEX), Polish investigators found that in the central parts of the Bransfield Strait, areas of dense krill concentration exhibited low chlorophyll $\underline{a}$ values at the surface (~ 0.5 mg/m^3) and in the water column (< 50 mg/m^2) (Lipski, personal commuication). Chilean scientists, also during the FIBEX, reported similar findings. Thus in the region to the south of the South Shetland Islands, Uribe (1982) found low values of chlorophyll $\underline{a}$ (< 0.5 mg/m^3) in the central part of the Bransfield Strait where high krill concentration in the 10-200 m layer was found. According to Uribe (1982) the poverty of the phytoplankton was not due to nutrient limitation (phosphate: 1.4 µg.at/l; nitrate: 17.9 µg.at/l), but most likely to intensive krill feeding. Further, the data from the R/V MELVILLE cruise (also during FIBEX) demonstrate on a temporal basis the inverse relationship often noted between phytoplankton biomass and zooplankton (mainly krill) abundance (Holm-Hansen and Huntley, in press).

IV. DISCUSSION

The general picture that emerges from the above investigations is the great variability of the phytoplankton biomass and primary productivity between low values typical of oligotrophic waters and high values characteristic of eutrophic regions. This geographical variability (up to two orders of magnitude) tends to overshadow the expression of seasonal differences. The two most significant findings resulting from these investigations are: (a) that there is much greater variability in the productivity parameters studied than had been previously thought, and (b) that the productivity of the Southern Ocean as a whole, perhaps, is not as high as we were led to believe.

With regard to the former, Fogg (1977) contends that the spatial variations in primary production in the waters south of the Polar Front cannot be accounted for by differences in incident radiation, water temperature, or concentration of nitrate, phosphate, and silicate, as each of these parameters has similar values over the entire area. A similar conclusion was reached by Holm-Hansen et al. (1977) who found it difficult or impossible to deduce rate-limiting factors by the direct comparison of any one parameter (e.g. temperature) as many factors vary simultaneously, including the species composition of phytoplankton (where significant floral changes take place at the Polar Front region). It would seem, however, that the physical structure of the water column (i.e. stability and the depth of the mixed layer), the near freezing temperatures of surface waters, and grazing are the most significant factors controlling phytoplankton production in the Southern Ocean.

The second significant finding -- namely that recently acquired data show that the primary productivity of the Southern Ocean is not as high as originally believed -- has far-reaching implications with regard to the future exploitation of the living resources of that ocean. Management decisions regarding the exploitation of these resources must be based on good estimates of primary productivity and on a better understanding of the flow of energy through the Antarctic marine ecosystem. A review of recent studies of Antarctic bacterioplankton, dissolved organic matter, nanoplankton and sea ice-algae are forcing us to to examine the classical description of the simple food chain from diatoms ⟶krill ⟶ whales. These studies suggest the presence of other pathways through which a major part of the available energy may be flowing. This new paradigm (Fig. 4) may contain yet other strands, so that the classic pathway may constitute only a part of the energy flow within the Antarctic marine ecosystem.

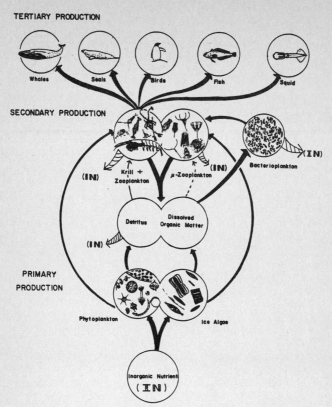

Fig. 4. Proposed paradigm of energy flow within the Antarctic marine ecosystem.

V. REFERENCES

Allanson BR, Hart RC and Lutjeharms JRE (1981) Observations on the nutrients, chlorophyll and primary production of the Southern Ocean south of Africa. South Afr J Antarct Res 10: 3-14.

von Bodungen B, Tilzer MM and Zeitzschel B (1982) Phytoplankton growth dynamics during spring blooms in Antarctic waters. Joint Oceanographic Assembly, 2-13 August 1982, Halifax, Canada, Abstracts of Invited Papers, p 59.

Braarud T and Klem A (1931) Hydrographical and chemical investigations in the coastal waters off Møre and in the Romsdalfjord. Hvalraadets Skrifter 1: 1-88.

Carlucci AF and Cuhel RL (1977) Vitamins in the south polar seas: distribution and significance of dissolved and particulate vitamin B_{12}, thiamine and biotin in the southern Indian Ocean. In: Llano GA (ed.) Adaptations within Antarctic Ecosystem, Proceedings of Third SCAR Symposium on Antarctic Biology. pp 115-128.

Doty MS and Oguri M (1956) The island mass effect. J Cons, Cons perma Int Explor Mer 22: 33-37.

El-Sayed SZ, Mandelli EF and Sugimura Y (1964) Primary organic production in the Drake Passage and Bransfield Strait. In: Lee MD (ed.) Biology of the Antarctic Seas I, American Geophysical Union, 1: 1-110.

El-Sayed SZ and Mandelli EF (1965) Primary production and standing crop of phytoplankton in the Weddell Sea and Drake Passage. In: Llano GA (ed.) Biology of the Antarctic Sea II, American Geophysical Union, 5: 87-106.

El-Sayed SZ (1967) On the productivity of the Southwest Atlantic Ocean and the waters west of the Antarctic Peninsula, Antarctic Research Series. In: Schmitt W and Llano G (eds.) Biology of the Antarctic Seas II. American Geophysical Union, 11: 15-47.

El-Sayed SZ (1968) Primary productivity of the Antarctic and Subantarctic. In: Bushnell V (ed.) Primary Productivity and Benthic Marine Algae of the Antarctic and Subantarctic. Folio 10, Antarctic Map Folio Series. American Geographical Society, p 1-6.

El-Sayed SZ (1970) On the productivity of the Southern Ocean. In: Holdgate MW (ed.) Antarctic Ecology Vol. I. Academic Press, New York, pp 119-135.

El-Sayed SZ (1971) Observations on phytoplankton bloom in the Weddell Sea. In: Llano GA and Wallen IE (eds.) Biology of the Antarctic Seas IV. American Geophysical Union, 17: 301-312.

El-Sayed SZ and Jitts HR (1973) Phytoplankton production in the southeastern Indian Ocean. In: Zeitschel B (ed.) The Biology of the Indian Ocean Vol. 3, Springer-Verlag, New York, pp 131-142.

El-Sayed SZ and Taguchi S (1981) Primary production and standing crop of phytoplankton along the ice-edge in the Weddell Sea. Deep-Sea Res 28: 1017-1032.

El-Sayed SZ and Turner JT (1977) Productivity of the Antarctic and tropical subtropical regions: A comparative study. In: Dunbar MJ (ed.) Proceedings of SCOR/SCAR Polar Oceans Conference, Montreal Canada, May, 1974. Arctic Institute of North America, pp 463-504.

El-Sayed SZ and Weber LH (1982) Spatial and temporal variations in phytoplankton biomass and primary productivity in the Southwest Atlantic and the Scotia Sea. Polar Biol 1: 83-90.

El-Sayed SZ, Stockwell DA, Reheim HR, Taguchi S and Meyer MA (1979) On the productivity of the southwestern Indian Ocean. In: Arnaud PM and Hureau JC (eds.) CNFRA, Campagne oceanographique MD 08/Benthos aux iles Crozet, Marion et Prince Edward: premiers resultats scientifiques, pp 83-110.

El-Sayed SZ, Holm-Hansen O and Biggs DC (1983) Phytoplankton standing crop, primary productivity, and near-surface nitrogenous nutrient fields in the Ross Sea, Antarctica. Deep-Sea Res 30(8A): 871-886.

Eppley RW (1972) Temperature and phytoplankton growth in the sea. Fish Bull 70: 1063-1085.

Fogg GE (1977) Aquatic primary production in the Antarctic. Phil Trans R Soc London B 279: 27-38.

Fryxell GA and Hasle GR (1971) Corethron criophilum Castracani: its distribution and structure. In: Llano GA and Wallen IE (eds.) Biology of the Antarctic Seas IV, Antarctic Research Series. American Geophysical Union, 17: 335-346.

Fukuchi J (1980) Phytoplankton chlorophyll stocks in the Antarctic Ocean. J Oceanogr Soc Japan 36: 73-84.

Gran HH (1931) On the conditions for the production of plankton in the sea. J Cons, Cons Perma Int Explor Mer 75: 37-46.

Hart TJ (1934) On the phytoplankton of the southwest Atlantic and the Bellingshausen Sea. Discovery Rept 8: 1-268.

Hart TJ (1942) Phytoplankton periodicity in Antarctic waters. Discovery Rept 21: 261-356.

Hasle GR (1969) An analysis of the phytoplankton of the Pacific Southern Ocean: abundance, composition and distribution during the Brategg Expedition 1947/48. Hvalard Skr 52: 1-168.

Holdgate MW (1970) Plankton and its pelagic consumers. In: Holdgate MW (ed.) Antarctic ecology Vol. 1, Academic Press, London New York, p 117.

Holm-Hansen O, El-Sayed SZ, Franceschini GA and Cuhel K (1977) Primary production and the factors controlling phytoplankton growth in the Antarctic seas. In: Llano GA (ed.) Adaptations within Antarctic Ecosystem, Proceedings of SCAR Symposium in Antarctic Biology. p 11-50.

Holm-Hansen O and Huntley M (in press) Feeding requirements of krill in relation to food source. J Crust Biol.

Horne AJ, Fogg GE and Eagle DJ (1969) Studies in situ of the primary production of an area of inshore antarctic sea. J Mar Biol Ass U K 49: 393-405.

Ichimura S and Fukushima H (1963) On the chlorophyll content in the surface water of the Indian and Antarctic Oceans. Bot Mag Tokyo 76: 395-399.

Jacobs SS and Amos A (1967) Physical and chemical oceanographic Observations in the Southern Ocean. Tech Report 1-Cu-1-67. Lamont-Doherty Geological Observatory. pp 287.

Jacques G (1983) Some ecophysiological aspects of the Antarctic phytoplankton. Polar Biol 2: 27-33.

Jacques G and Minas M (1981) Production primaire dans le secteur indien de l'Ocean Antartique en fin d'eté. Oceanol Acta 4: 33-41.

Kennett JP (1977) Cenozoic Evolution of antarctic glaciation, the circum-Antarctic Ocean, and their impact on Global Paleoceanography. J of Geoph Res 82: 3843-3859.

Klyashtorin LB (1961) Pervichnaya produktsiya v Atlanticheskon i Yuzhnom Okranakh po dannym pyatogo Antarkticheskogo reisa dizel-elektrokhoda. Ob Dokl Akad Nauk SSSR 141: 1204-1207

Mandelli EF and Burkholder PR (1966) Primary productivity in the Gerlache and Bransfield Straits of Antarctica. J Mar Res 24: 15-27.

Marshall PT (1957) Primary production in the Arctic. J Cons, Cons Perma Int Explor Mer 23: 173-177.

Neori A and Holm-Hansen O (1982) Effect of temperature on rate of photosynthesis in Antarctic phytoplankton. Polar Biol 1: 33-38.

Olson RJ (1980) Nitrate and ammonium uptake in Antarctic waters. Limnol Oceanogr 25: 1064-1074.

Pingree RD (1978) Cyclonic eddies and cross frontal mixing. J Mar Biol Assoc U K 58: 955-963.

Richards FA and Thompson TG (1952) The estimation and characterization of plankton populations by pigment analysis. 2, A spectrophotometer method for estimation of plankton pigment. J Mar Res 11: 156-172.

Saijo Y and Kawashima T (1964) Primary production in the Antarctic Ocean. J Oceanogr Soc Japan 19: 190-196.

Sverdrup HU (1953) On conditions for the vernal blooming of phytoplankton. J Cons, Cons Perma Int Explor Mer 18: 287-295.

Uribe E (1982) Influence of the phytoplankton and primary production of the Antarctic waters in relationship with the distribution and behavior of krill. Inst Antart Chil Scientific Series, No. 28 pp 147-163.

Walsh JJ (1971) Relative importance of habitat variable in predicting distribution of phytoplankton at the ecotone of the Antarctic upwelling ecosystem. Ecol Monogr 41: 291-309.

Yamaguchi Y and Shibata Y (1982) Standing stock and distribution of phytoplankton chlorophyll in the Southern Ocean south of Australia. Trans Tokyo Univ Fish, No. 5 pp 111-128.

A THERMODYNAMIC DESCRIPTION OF PHYTOPLANKTON GROWTH

D.A. KIEFER

Department of Biological Sciences, University of Southern California, Los Angeles, California 90089, USA

I. INTRODUCTION

As early as 1942 Monod noted that the growth rate for a continuous culture of bacteria was adequately described by a hyperbolic function. Since Monod's work, there have been numerous observations of similarities in the growth kinetics of microbes and the kinetics of reactions catalyzed by enzymes in vitro (e.g. Malek and Fencl,1966). Such similarities have suggested to researchers that the growth response of microbes may be simply explained by the kinetics of a rate limiting step in a metabolic pathway. Thus, parameters of Michaelis-Menten kinetics, the half saturation constant and the maximum velocity for substrate-saturated systems, have been adapted for descriptions of the microbial growth. In addition, this kinetic description of growth has served as the conceptual basis for the establishment of the steady state in continuous culture devices.

While the kinetic description of enzyme action has been usefully applied to a description of phytoplankton growth (eg. Caperon,1965; Droop,1968), its application offers little insight into the regulation of metabolism. This limitation has been the stimulus to explore an alternate description of steady state microbial growth, based upon principles of non-equilibrium thermodynamics (Kiefer and Enns,1976). In this model the biochemical reactions occurring during the conversion of light energy into chemical bond energy are characterized by two pairs of coupled reactions. Both pairs of reactions are described phenomenologically and are assigned a degree of coupling. In the present paper I will describe the model and apply it to a description of light-limited and nutrient-limited growth of phytoplankton. These predictions will then be compared with studies of general metabolic regulation by phytoplankton grown in continuous culture.

II. DESCRIPTION OF MODEL

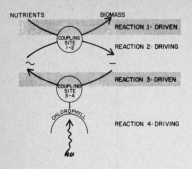

Figure 1. A simplified representation of the reactions leading to the growth of phytoplankton. The driving reactions release free energy while the driven reactions consume free energy.

Figure 1 presents the principal biochemical reactions that are to be described mathematically. The conversion of light and nutrients into cellular material involves two sets of coupled reactions which are connected by a cycling photosynthetic electron transport system, symbolized by:

$$ \quad (1)$$

At coupling site 3-4 the free energy of absorbed quanta reaching reaction centers (reaction 4), drives the separation of charge between redox pools of electron donors and acceptors (reaction 3). For the coupling site 1-2 the free energy released in photosynthetic electron flow (reaction 2) drives the anabolic reactions of growth (reaction 1), which may be represented by:

$$6.14 CO_2 + 3.75_2O + NH_3 = C_{6.14}H_{10.4}O_{2.24}N + 6.90_2 \quad \text{(Kok, 1960)} \quad (2)$$
$$\Delta \bar{G} = 140 \text{ kcal} \cdot \text{mole } O_2^{-1}$$

The series of equations describing the flow of quanta and electrons through the system shown in Figure 1 are:

$$J_4 = 1.3 \times 10^5 \cdot J_i \cdot \text{Chl a} \quad (3)$$

$$J_4 = L_{34}A_3 + L_{44}A_4, \qquad A_4 = +40.0 \quad (4)$$

$$J_3 = L_{33}A_3 + L_{34}A_3, \quad (5)$$

$$J_2 = L_{12}A_1 + L_{22}A_2, \qquad A_1 = -17.5 \quad (6)$$

$$J_1 = L_{11}A_1 + L_{12}A_2, \quad (7)$$

$$J_2 = J_3, \qquad A_2 = A_3 \quad (8), (9)$$

$$q_{12} = \frac{L_{12}}{\sqrt{L_{11}L_{22}}} = 0.91 \tag{10}$$

$$z_{12} = \frac{\sqrt{L_{11}}}{\sqrt{L_{22}}} = 0.96 \tag{11}$$

$$q_{34} = \frac{L_{34}}{\sqrt{L_{33}L_{44}}} = 0.95 \tag{12}$$

$$z_{34} = \frac{\sqrt{L_{33}}}{\sqrt{L_{44}}} = 1.00 \tag{13}$$

$$A_3 = f(L_{11}...L_{44}, A_1, A_4) \tag{14}$$

Equation 3 describes the flux of quanta (white light) absorbed by a thin suspension of phytoplankton cells. J_4, the flux of absorbed quanta, is a function of the incident level of irradiance, J_i, and the chlorophyll concentration in the suspension, Chl. The constant in equation 3 is the product of the molar absorption coefficient for cellular chlorophyll and a volume element which relates the incident level of irradiance to a one liter suspension. Equations 4-7 are phenomenological equations for the two pairs of coupled reactions shown in Figure 5. J_4 is the flux of absorbed quanta, J_3 is the rate of electron flow across reaction centers, J_2 is the rate of photosynthetic electron flow, and J_1 is the flux of electrons to carbon dioxide. The stoichiometry of equation 2 indicates that 4.5 moles of electrons are transferred for each mole of carbon dioxide assimilated. A_1 through A_4 are the chemical affinities for the four reactions. In the case of light limited metabolism, A_1 and A_4 are constant, but A_2 and A_3 may have any value consistent with the constraints placed upon the system. The L terms of these equations are the phenomenological coefficients, which index the conductivity or catalytic capacity of the metabolic pathways. Equations 8 and 9 are conditions for cycling of reactions 2 and 3.

Equations 10-13 add additional constraints to the phenomenological coefficients and represent a major assumption of the model. Use of these equations requires that the degrees of coupling q_{12} and q_{34} and the ratios of straight coefficients z_{12} and z_{34} are constant at all growth rates. Values for these parameters have been calculated from studies of the growth of <u>Chlorella</u>.

Equation 14, which is a complex function, restricts A_2 to a value which optimizes the overall efficiency of energy conversion by the system. For given values of A_1, A_4, q_{12}, q_{34}, z_{12}, and z_{34}, there is an optimum value for A_2. If A_2 decreases below $+28$ kcal·(mole e)$^{-1}$, the overall efficiency of energy conversion, n_{14}, drops below its maximum value of 0.23. This decrease is caused mostly by a rapid decrease in the efficiency at coupling site 1-2. On the other hand, if A_2 increases above a value of $+28$, the overall efficiency of energy conversion drops because of a rapid decrease in the efficiency at coupling site 3-4.

III. PREDICTIONS FOR LIGHT LIMITED GROWTH

Since A_1 and A_4 are known constants and J_i is an input into the model, the 12 equations (3-14) contain 13 unknowns. An additional relationship, 15a, describes the chlorophyll content of the phytoplankton cells as a function of incident light intensity.

$$Chl = 6.0 \times 10^{-4} - 1.8 \times 10^{-2} \cdot J_i \quad (15a)$$

when $J_i \leq 1.9 \times 10^{-2}$ m ein·cm^{-2}·hr^{-1}

This empirical relationship is based upon measurements of the light-limited continuous culture of <u>Chlorella pyrenoidosa</u> by Myers and Graham (1971); the observed linear relationship between cellular chlorophyll and light intensity is limited to intensities that are subsaturating to growth.

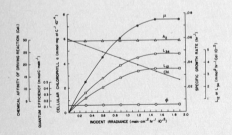

Figure 2. Responses to light levels predicted by the thermodynamic model. This figure shows variations in specific growth rate, μ, cellular chlorophyll, Chl, catalytic capacity at the 2 coupling sites, L_{12} and L_{34}, quantum efficiency, ϕ, and the chemical affinity of reaction 2, A_2, with variations in irradiance.

(The solution of equations 3-15a at subsaturating light levels are presented in Figure 2.)

According to the model, the cells maintain a constant quantum efficiency of photosynthesis, $\phi = J_{CO_2}/J_4$, and a constant driving potential for photosynthetic electron transport despite changes in growth rate and cellular chlorophyll. Changes in the cross coefficients, L_{12} and L_{34} of the phenomenological equations parallel changes in growth rate. As discussed previously, these conductivity terms index the enzymatic or catalytic capacity of the cells. In light-limited growth where the concentration of nutrients remains constant, decreases in L_{12} and L_{34} with decreases in irradiance reflect decreases in either enzyme concentration or specific activity at both coupling sites.

Looking at Figure 1, we summarize the metabolic regulation by phytoplankton required for light-limited growth as follows. A decrease in irradiance will slow reaction 4, which will slow reaction 3 because they are tightly coupled. A decrease in reaction 3 will lead to a transient decrease in the chemical affinity of reaction 2. In order to reestablish the chemical affinity required for optimal efficiency of energy conversion, the rate of reactions 1 and 2 must decrease with respect to reactions 3 and 4. This is achieved by decreasing the conductivity at both coupling sites and increasing the chlorophyll content of the cells. Such alterations in enzymatic activity and capacity for gathering radiant energy are consistent with the observations on <u>Chlorella</u> by Myers and Graham (1971). The observations are also consistent with Björkman's (1968) observation that the enzyme ribulose 1,5-diphosphate carboxylase is found in much lower concentration in shade plants than in sun plants.

IV. PREDICTIONS FOR CARBON-LIMITED GROWTH

According to the phenomenological equations (eqs. 4-7), fluxes are affected by either variations in the phenomenological coefficients or by variations in chemical affinities. In the system described by Figure 1, decreases in CO_2 concentration may cause a decrease in the chemical affinity of reaction 1 (see equation 2) or a decrease in phenomenological coefficients, L_{11} and L_{12}. The decrease in coefficients is a much more important factor in regulation than the decrease in chemical affinity.

In order to describe carbon-limited growth, we again invoke equations 3-14, which describe light absorption, electron flow, constant degrees of coupling, and optimal efficiency in energy conversion. The empirically-derived relationship, 15a, is however replaced by:

$$L_{12} = \frac{5.2 \times 10^{-2} (CO_2)}{5.0 \times 10^{-3} + (CO_2)} \tag{15b}$$

This equation describes changes in the phenomenological cross coefficient with changes in carbon dioxide concentration. It is assumed that the enzyme ribulose 1,5-diphosphate carboxylase is the rate limiting enzyme for carbon-limited growth, and that variations in L_{12} are caused by an insufficient concentration of CO_2 to saturate the enzyme. The equation is of the form of the Michaelis-Menten equation for enzyme kinetics with L_{12} substituted for V, the velocity of the catalyzed reaction. The constant 5.2×10^{-2} is the maximum value of L_{12} when the enzyme is saturated, and the constant 5.0×10^{-3} is the half-saturation constant for the enzyme (Bahr and Jensen, 1974). The constant 5.2×10^{-2} was obtained by solving the equation for standard growth conditions when the concentration of CO_2 was 10 μm.

Since the chemical affinity of reaction 4, A_4, is known, the 13 equations of the CO_2-limited model may be solved for varying concentrations of CO_2; the results are shown in Figure 3. The response of the model cells to limiting concentrations of CO_2 is approximately hyperbolic and unlike light-limited growth, the cellular concentration of chlorophyll decreases with growth rate. Both the quantum efficiency of photosynthesis and the chemical affinity for photosynthetic electron transport vary little, as they did in light-limited growth. The decrease in the phenomenological cross coefficient, L_{12}, is an input into the model and is caused by a decrease in substrate concentration. The concentration and specific activity of the rate limiting enzyme, ribulose 1,5-diphosphate carboxylase, will remain constant. The cross coefficient for coupling site 3-4, L_{34}, decreases in parallel with chlorophyll content and specific growth rate.

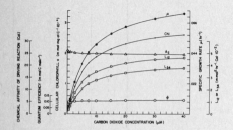

Figure 3. Responses to the concentration of carbon dioxide predicted by the thermodynamic model. The responses include specific growth rate, μ, cellular chlorophyll, Chl, catalytic capacity at the two coupling sites, L_{12} and L_{34}, quantum efficiency, φ, and the chemical affinity of reaction 2, A_2.

Referring again to Figure 1, we may summarize carbon-limited regulation as follows. A decrease in carbon dioxide concentration will slow reaction 1 because of a decreased conductivity. Since reactions 1 and 2 are coupled, reaction 2 will decrease, leading to a transient increase in the chemical affinity of reaction 2. In order to reestablish the chemical affinity required for optimal efficiency of energy conversion, the rate of coupled reactions 3-4 must decrease. This decrease is achieved by decreases in both the chlorophyll content of the cells and the conductivity at coupling site 3-4. Thus, cells in the new steady state growth have a lower content of chlorophyll and lower levels of enzyme activity. Although there are too few measurements of carbon-limited growth to test many of these predictions, studies of steady-state growth limited by other macronutrients are consistent with the predictions of the carbon-limited model.

If the thermodynamic model is of value, its predictions must be consistent with measurements of growth in continuous culture, and the model should suggest general mathematical relationships that can be usefully applied to descriptions of phytoplankton growth. The concluding section of this paper will include not only a comparison of model predictions with measurements but also the derivation of a general mathematical description of steady state growth.

V. STEADY STATE GROWTH IN CULTURES

The predictions of the thermodynamic model have been compared with measurements (by Laws and Bannister, 1980) of the growth in continuous culture of the marine diatom, <u>Thalassiosira weissflogii</u>. Cultures were grown in either chemostats where the rates of supply of either ammonium, nitrate, or phosphate were limiting to growth or in turbidostats where light levels were limiting. In the chemostats, light levels were both constant and saturating for growth; in the turbidostats nutrient concentrations were saturating. The results of such studies are conveniently summarized in Figure 4, in which the specific growth rate of the culture is plotted against the ratio of cellular chlorophyll to carbon.

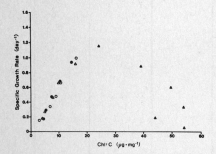

Figure 4. The specific growth rate of <u>Thalassiosira weissflogii</u> and corresponding ratios of cellular chlorophyll and carbon. Data are from Table 1 of Laws and Bannister (1980), and symbols for the different limitations are O - nitrate, ⊙ - ammonium; ▲ - phosphate, and △ - light.

It appears from this figure that the response of the diatoms to variations in either of the three limiting nutrients are similar and that this response is distinct from that for light-limitations. When the predictions of the models shown in Figures 2 and 3 are plotted accordingly, a similar pattern exists.

The conversion of light energy into cellular material can be described as merely the product of the flux of light absorbed by the cells and the quantum efficiency of carbon assimilation (eg. see Bannister,1974):

$$\mu + r = \phi \cdot {}^{Chl}/C \cdot {}^{o}a_p \cdot E_o \tag{16}$$

μ, r, ϕ, ${}^{o}a_p$, E_o are respectively the specific growth rate (units day^{-1}), the specific respiration rate (units of day^{-1}), the quantum efficiency (moles C·Einst^{-1}), the specific absorption coefficient (m^2·mg Chl^{-1}), and the incident scalar irradiance (Einst·m^{-2}·day^{-1}). Referring to Figures 2 and 3, one sees that the thermodynamic model predicts that for both light- and nutrient-limited growth the ratio of chlorophyll to carbon, $^{Chl}/C$ will vary with growth rate while ϕ will remain constant. If this prediction is true and if ${}^{o}a$ also remains constant, equation 16 indicates that a plot of $\mu+r$ versus the product $(^{Chl}/C)E_o$ for both light- and nutrient-limited growth will yield a linear relationship whose slope is equal to the product of the two constants, $\phi^{o}a$. The data shown in Figure 4 was replotted accordingly and a merging of the light-limited and nutrient-limited indeed occurred (Kiefer and Mitchell, 1983).

When the plot of $\mu+r$ versus $(^{Chl}/C)E_o$ was examined, it was apparent that while the relationship was linear for nutrient limited growth, there was some curvature in the light-limited data. At low values of $(^{Chl}/C)E_o$ the slope is larger than for high values. Such nonlinearity is likely to be caused by decreases in the values of either ϕ or ${}^{o}a_p$ with increased irradiance. Since Laws and Bannister did not measure ${}^{o}a_p$, we cannot distinguish between changes in the two "constants" and are forced to assume that the nonlinearity is due primarily to decreases in ϕ as light levels become saturating to growth.

Figure 5 shows calculated values for ϕ at varying light levels for the light-limited cultures of T. weissflogii. The values were calculated by introducing measured values of μ, r, $^{Chl}/C$, and E_o and an assumed value of 17 m^2·gm Chl^{-1} for ${}^{o}a_p$ (Bannister,1979) in equation 16. A systematic decrease in ϕ as light intensities approach saturating levels is evident in the figure. As shown in Figure 5 we have found that the function $\phi(E_o)$ is described well by:

$$\phi(E_o) = \frac{\phi_m K_\phi}{K_\phi + E_o} \tag{17}$$

ϕ_m and K_ϕ are constants, ϕ_m being the maximum quantum efficiency and K_ϕ being the irradiance at which the quantum efficiency is equal to $\phi_m/2$. Figure 5 includes a graph of equation 17 in which ϕ_m has a value of 0.60 g-atom C·Einst^{-1} and K_ϕ is 10 Einst·m^{-2}·day^{-1}.

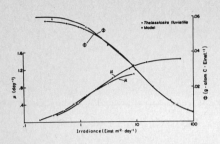

Figure 5. Predicted quantum efficiency and specific growth rate of light-limited cells. The other pair of curves (X) show measured quantum efficiencies and specific growth rates for Thalassiosira weissflogii, growing under light limitations.

Figure 5 also includes calculations of the light-limited growth rates obtained by substituting $\phi(E_o)$ (equation 17) for ϕ in equation 16 and calculating $\mu+r$ for differing values of E_o and $^{Chl}/C$. Values for μ were then obtained from $\mu+r$ by application of the empirical equation of Laws and Bannister (1980).

Equation 17 is substituted for ϕ in equation 16 to yield:

$$\mu+r = \frac{\phi_m K_\phi \text{ Chl } a_p E_o}{C(K_\phi + E_o)} \tag{18}$$

It is proposed that this equation provides a general description of the steady state growth of phytoplankton that is consistent with the behavior of the thermodynamic model. As a test of the validity of this equation, we have introduced measured values of E_o, C, and Chl for light- and nutrient-limited growth of T. weissflogii into equation 18. The predicted values of $\mu+r$ were then compared with the values of growth and respiration measured by Laws and Bannister as shown in Figure 6. It is evident from this figure that equation 18 both linearizes the data shown in Figure 4 and accurately predicts the growth rate of T. weissflogii in continuous culture.

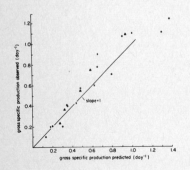

Figure 6. Observed and corresponding predicted gross production for Thalassiosira weissflogii that are limited in growth rate by either nutrients or light. Predicted rates were calculated from equation 18. Symbols for different limitations are: X - nitrate; 0 - ammonium; Δ - phosphate; 0 - light.

Acknowledgments

This work was supported by several agencies: the National Oceanic and Atmospheric Administration, Grant 04-7-158-44123, the National Aeronautics and Space Administration, Grant NAGW-317, and the Office of Naval Research, Contract N00014-81-K-0388.

VI. REFERENCES

Bahr, JT and RG Jensen (1974) Ribulose diphosphate carboxylase from freshly ruptured spinach chloroplasts having an in vivo $K_m[CO_2]$. Plant Physiol. 53:39-44.

Bannister, TT (1974) Production equations in terms of chlorophyll concentration, quantum yield, and upper limit to production. Limnol. Oceanogr. 19:1.

Bannister, TT (1979) Quantitative description of steady state, nutrient-saturated algal growth, including adaptation. Limnol. Oceanogr. 24:76-96.

Björkman, O (1968) Carboxydismutase activity in shade-adapted and sun-adapted species of higher plants. Physiologia Plantarum 21:1-10.

Caperon, J (1965) The dynamics of nitrate-limited growth of Isochrysis galbana populations. Ph.D. Thesis, Univ. of California, San Diego.

Crofts, AR, CR Wraight and DE Fleischman (1971) Energy conservation in the photochemical reactions of photosynthesis and its relation to delayed fluorescence. F.E.B.S. Letters 15:89.

Droop, MR (1968) Vitamin B-12 and marine ecology. IV. The kinetics of uptake, growth, and inhibition in Monochrysis lutheri. J. Mar. Biol. Assoc. U.K. 48:689.

Kiefer, DA and T Enns (1976) A steady-state model of light-, temperature;, and carbon-limited growth of phytoplankton. In: RP Canale (ed.), Modeling Biochemical Processes in Aquatic Ecosystems. Ann Arbor Science, Ann Arbor, Michigan. pp 319-336.

Kiefer, DA and BG Mitchell (1983) A simple, steady-state description of phytoplankton growth based on absorption cross section and quantum efficiency. Limnol. Oceanogr. 28:770-776.

Kok, B (1952) On the efficiency of Chlorella growth. Acta Botanica Neerl I, 445.

Laws, EA and TT Bannister (1980) Nutrient- and light-limited growth of Thalassiosira fluviatilis in continuous culture, with implications for phytoplankton growth in the ocean. Limnol. Oceanogr. 25:457-473.

Malek, I and Z Fencl (1966) Theoretical and Methodological Basis of Continuous Culture of Microorganisms. Academic Press: New York and London.

Monod, J (1942) La croissance des cultures bacteriennes. Herman: Paris.

Myers, J and J Graham (1971) The photosynthetic unit in Chlorella measured by repetitive short flashes. Plant Physiol. 48:282-286.

MECHANISMS OF ORGANIC MATTER UTILIZATION BY MARINE BACTERIOPLANKTON

F. AZAM and J.W. AMMERMAN

Institute of Marine Resources, Scripps Institution of Oceanography, La Jolla, California 92093, USA.

Recent work indicates that bacteria in the sea may consume 10-50% of the primary production (Hagstrom et al., 1979; Fuhrman and Azam, 1980, 1982; Williams, 1981), and the resulting bacterial secondary production is presumably channelled into a dynamic microbial foodweb (Azam et al., 1983). These observations have generated much interest in the coupling between the production of organic matter and its utilization by heterotrophic bacterioplankton, since this coupling may determine how the primary production is partitioned between the classical grazing food chain and the emergent microheterotrophic foodweb.

The finding that bacterioplankton is a major route for the flow of matter and energy in the marine foodweb is in apparent contradiction with the conventional wisdom that bacterial nutrients are too dilute in seawater to support rapid bacterial growth. In this paper, we will argue that marine heterotrophic bacteria utilize organic matter in close metabolic and spatial coupling with the sources of organic matter, and this explains why they are so effective in consuming organics from seawater.

Sources and Modes of Organic Matter Production

Bacterial nutrients in seawater are chemically and physically diverse. The dissolved organic matter (DOM) contains, at very low concentrations, many solutes which are directly utilizable by marine bacteria (e.g. sugars, amino acids, organic acids). Some other solutes (e.g. proteins, polysaccharides, organophosphorus compounds) are not transported into bacteria; they are generally hydrolyzed extracellularly and the hydrolysis products transported across the palsmalemma of the bacterial cells. Likewise, particulate organic matter (POM) must be hydrolyzed to make it accessible to membrane transport systems.

Phytoplankton organic matter becomes available for bacterial uptake by a number of mechanisms, either directly from algae or from herbivore feces (Williams, 1981). Healthy algae apparently exude a certain (variable) fraction of the photosynthate, although the extent and the nature of algal exudation is somewhat controversial (Sharp, 1977). Metabolic pools of algae might at times be released into seawater during predation on the alga by a herbivore (Lampert, 1978). Senescent algae release DOM into seawater via leakage and autolysis. Finally, animal feces and excretions introduce DOM into seawater. The relative contribution of these processes to DOM input is not known with certainty. It is important to note however that all DOM sources are particulate, hence DOM production events would create microzones of high DOM concentrations in the vicinity of the source. The non-random distribution of nutrient molecules may have significant implications for nutrient uptake kinetics of bacteria. Given the above scenario of DOM production, how do bacteria in the sea metabolically interact with the organic matter in their environment to optimize nutrient uptake and growth?

Uptake Kinetics of Directly Utilizable Compounds

It has been generally thought that planktonic marine bacteria have evolved very high affinity membrane transport systems for the uptake of dissolved nutrients present in nanomolar concentrations in the seawater (Wright and Burnison, 1979). Azam and Hodson (1981) showed that assemblages of marine bacteria exhibited multiphasic uptake kinetics for D-glucose uptake, with K_m range of 10^{-9} to 10^{-4} M. They reasoned that high K_m (high V_{max}) uptake systems indicate the presence of microzones of high substrate concentration in a bacterium's microenvironment. Nissen et al. (in press) studied glucose uptake kinetics in a marine bacterial isolate, LNB-155. They found evidence of a single transport system for glucose uptake which changed its K_m (range: 10^{-8} to 10^{-3} M) in an all-or-none fashion at certain critical concentrations of glucose in the medium. They suggest that a multiphasic transport system would provide metabolic flexibility for a bacterium living in a microenvironment where very low (nanomolar) to high substrate concentrations might be encountered. Since low K_m, low V_{max} systems saturate at low substrate concentrations, the presence of high K_m (and high V_{max}) transport systems allows enhanced rates of substrate uptake over a broad range of environmental substrate concentrations.

Multiphasic transport systems for ℓ-leucine persisted in an assemblage of marine bacteria which had undergone 30 divisions in a continuous culture fed with particle-free unenriched seawater (Hagstrom et al., in press). ($K_m + Sn$) varied from 3 nM to 230 nM (Sn = substrate concentration). The high K_m (and high V_{max}) systems are either constitutive or may be induced by high ℓ-leucine concentrations which could occur in the periplasmic space of gram-negative bacteria during protein hydrolysis (Hollibaugh and Azam, 1983). Considering the high metabolic cost of maintaining a transport system, it would appear that planktonic marine bacteria must frequently experience high ℓ-leucine concentrations in their microenvironment for such high K_m transport systems to be constitutive.

We asked whether the seawater DOM pool could support the growth of planktonic bacteria in the absence of POM. Using 0.2 μm Nuclepore filtered seawater as growth medium we were able to grow mixed assemblages of bacterioplankton in batch or continuous cultures at doubling times from 6h to 39h (Ammerman et al., in press; Hagstrom et al., in press). These growth rates are in the range observed for natural populations of marine bacteria (Fuhrman and Azam, 1980, 1982). It appears therefore that at least some bacteria in seawater are able to grow at the expense of the bulk-phase concentrations of DOM (and without exposure to high nutrient microzones around particulate sources of DOM). Growth in enriched microzones could be more rapid.

It appears that we need to revise our views regarding seawater DOM as a medium for bacterial growth. Clearly, seawater DOM is not too dilute to support bacterial growth; it is the bacteria that maintain the utilizable DOM (UDOM) components at the exceedingly low levels found in seawater. New inputs of nutrients, possibly at high concentrations in a production microzone, are effectively utilized by high K_m, high V_{max} uptake systems. The nutrients which do diffuse into the bulk-phase are scavenged by the high affinity low capacity systems. The result is a tight coupling between the production of UDOM and its uptake by bacteria, so these compounds do not accumulate. This view is supported by the observation that DFAA levels in the euphotic zone vary within an order of magnitude during a diel cycle despite rapid flux through the DFAA pool (Mopper and Lindroth, 1982). In another study (Ammerman and Azam, 1981) diel excursions of the 3',5'-cyclic AMP (cAMP) concentration in seawater (in the picomolar range) were apprently kept in check by an extremely high affinity transport system with a K_m of a few picomolar.

Polymer Utilization

Dissolved proteins and polysaccharides in the euphotic zone of the sea probably represent a quantitatively important route for matter and energy flux into bacterioplankton. It is of interest, therefore, to know the dynamics of protein and polysaccharide pools, and how bacteria utilize these polymers. In one study, concentration of dissolved protein off Scripps pier was found to be 950 nM (as glycine equivalents) while the concentration of dissolved free amino acids (DFAA) was only 23 nM (Hagstrom et al., in press). Polysaccharide concentrations are generally even higher than protein. Off Scripps pier we found 2.7 µM total carbohydrate, mostly polymeric (Hagstrom et al., in press). Similar values were reported by Burney et al. (1981). The monomeric sugars occur at much lower concentrations.

Recent work (Hollibaugh and Azam, 1983) found little dissolved exoprotease activity in coastal surface waters; most of the activity was associated with bacterial cells. The high metabolic cost of maintaining significant dissolved protease activity may favor hydrolysis by cell-associated proteases. The amino acid molecules produced by protein hydrolysis were taken up by bacteria in preference to those in seawater suggesting a close metabolic coupling between protein hydrolysis and amino acid uptake. Protein hydrolysis probably occurs in the periplasmic space of bacteria, creating high DFAA concentrations near the transport sites. Still, a significant fraction of the released amino acids diffuse into the seawater (Hollibaugh and Azam, 1983). It is likely that dissolved polysaccharides are utilized in a similar fashion, although this has not been studied for planktonic bacteria.

Utilization of Organophosphorus Compounds

Organophosphorus (OP) compounds generally do not traverse the cell membrane, the phosphate must first be cleaved by hydrolysis. We have found (Ammerman and Azam, unpublished) that a bacterial periplasmic or membrane bound enzyme, 5'-nucleotidase, may account for a large fraction of the OP hydrolysis in coastal waters. As in the case of exoproteases, most of the nucleotidase activity is also associated with the cells rather than free in seawater. Importantly, the Pi molecules released from 5'-nucleotides are 10-40 times more likely to be taken up by the microbial populations than the Pi molecules present in the bulk-phase seawater. This preference for uptake of hydrolysis products is strong evidence for a tight metabolic coupling between UDOM production and utilization. Although we have not tested it, we predict that bacteria also take up the organic moiety of the OP compound hydrolyzed in preference to the same molecular species in the bulk-phase.

Non-random Distribution of Bacterioplankton

We have so far discussed how marine bacteria might interact metabolically with DOM components diffusing up to the cell surface. It would seem that bacteria could enhance nutrient uptake by being in close vicinity of a source of DOM. We have recently proposed that bacteria might cluster in the vicinity of sources of sustained UDOM production (e.g. exuding algal cells or detritus undergoing hydrolysis) (cluster hypothesis; Azam and Ammerman, 1984). This spatial coupling would increase the encounter frequency between bacteria and the DOM molecules emanating from the source. We think that the clustering bacteria, due to the existence of high K_m, high V_{max} uptake systems, could remove a significant part of the new DOM before it diffuses into the bulk-phase. Such clustering could increase the degree of coupling between DOM production and its utilization by bacterioplankton.

Chemical Cues in Bacterial Algal Interactions

Healthy phytoplankton cells generally do not have any attached bacteria. Bacteria readily attach to dead algae and detritus and are believed to be important in POM decomposition. It is not clear why marine bacteria do not attack healthy phytoplankton, but the general belief is that algae excrete repellents to keep bacteria away (Sieburth, 1968). Yet, algal exudation is apparently one of the major sources of UDOM for bacteria (Hagstrom et al., 1979; Williams, 1981; Wiebe and Smith, 1977). Bacteria are attracted by one or more components of algal exudates (Bell and Mitchell, 1972). Smith and Higgens (1978) have suggested that metabolic feedback between bacteria and algae might exist. However, the identity of metabolic effectors (positive or negative) exchanged between bacteria and algae remains an exciting research problem.

It is interesting in this context that cAMP, a ubiquitous metabolic regulator, was found to be excreted by freshwater algae (Franko and Wetzel, 1980, 1981). Ammerman and Azam (1981) found dissolved cAMP in seawater samples in concentration range of 1 to 30 pM, and its concentration showed diel variations in a pattern similar to some bacterial nutrients. We found that marine bacterial assemblages transport cAMP by a highly specific, active transport system with a picomoplar K_m. This transport system is very effective in cAMP uptake at environmental concentrations, and can double the bacterial intracellular cAMP pool within minutes to hours (Ammerman and Azam, 1982).

We have speculated (Azam and Ammerman, 1984) that cAMP may act as a metabolic cue in algal-bacterial interactions. The characteristics of the cAMP transport system in bacteria indicate a regulatory rather than a nutritional role for cAMP. The cAMP transport system shows near-absolute specificity for cyclic nucleotides, and a preference for cAMP. AMP, which occurs at about 100 times greater concentration in seawater, is not a substrate of the cAMP transport system (Ammerman and Azam, 1982). It is highly unlikely that marine bacteria evolved the cAMP system to take up this compound as a nutrient. cAMP plays a key role in the regulation of catabolic enzyme synthesis in bacteria (Rickenberg, 1974). It is tempting to speculate that, if exuded by the algae along with bacterial nutrients, cAMP could serve as a unifying signal for the nutrient status of the microenvironment of the bacterium. However, we have no direct evidence for this speculation.

The fact that bacteria attack detritus but not healthy phytoplankton may have implications for the pattern of energy and matter flow in marine planktonic foodweb. If bacteria were to (hydrolytically?) attack healthy phytoplankton then they would be competing with the herbivores for algal POM, which they apparently do not. However, algal exudation makes some of the primary production metabolically accessible to bacterioplankton. It may be that an interplay of positive and negative effectors between bacteria and phytoplankton cells regulates algal exudation and leads to a close coupling between production and bacterial uptake of the exudates. Clustering of bacteria around the algal cells, but without attachment, may result from this interaction (Azam and Ammerman, 1984).

Consequences of being small and gram-negative

Planktonic marine bacteria are exceedingly small, much smaller than the enteric bacteria which have traditionally been studied by the bacteriologist. Most marine bacteria (in nature) are 0.2 - 0.6 μm in equivalent spherical diameter (Fuhrman, 1981), hence they have a very large surface to volume ratio. This means that marine bacteria, per unit biomass, have a large membrane area for the locationzation of the enzymes which hydrolyse polymers and the transport systems which take up the resulting monomers. This large surface to volume ratio in marine bacteria therefore should provide an intimate metabolic contact with the organic matter in the environment.

Most marine bacteria are gram-negative, and hence they have a periplasmic space. The periplasmic space is a region between the outer membrane and the plasmalemma of the gram-negative bacteria, containing a number of enzymes and solute binding proteins. The periplasmic enzymes have a function in hydrolysis of polymers and OP compounds, while the binding proteins, presumably free in the periplasmic space, may have a role in solute transport. The periplasmic space in enteric bacteria *Salmonella typhimurium* and *Escherichia coli* is estimated to occupy 20-40% of the cell volume (Stock et al., 1977). If the thickness of the periplasmic space in marine bacteria is the same as in the enterics then (because of the small diameter of marine bacteria) the periplasmic space might occupy an even greater percent of the cell volume. If these periplasmic proteins assist in nutrient aquisition, then gram-negative bacteria may be better adapted than gram-positive bacteria to the dilute and fluctuating nutrient levels encountered in seawater. Several interesting metabolic consequences result from the extremely small size of marine bacterioplankton, especially those at the lower end of the size spectrum. A 0.2 µm diameter bacterium has a cell volume of merely 0.0033 µm^3 or 3.33×10^{-18}ℓ. If this cell took up one molecule (1.66×10^{-24} moles) of an amino acid, the intracellular concentration of that amino acid would increase by 0.54×10^{-6}M, which is 100 to 1000-fold the concentration of individual amino acids in the bulk-phase seawater. When one considers that a significant fraction of the cell volume is occupied by the periplasmic space, then the increase in the intracellular concentration per molecule would be even greater than 0.54 µM. This means that uptake from bulk-phase seawater of solutes such as amino acids will always be against concentration gradients, and hence must be energy-dependent. It also follows that the uptake of a very few effector molecules per cell could profoundly affect the metabolic regulation. The intracellular concentration of cAMP is on the order of 1 µM (Ammerman and Azam, 1982 and references therein), which for the very small marine bacteria corresponds to only 1 or 2 molecules per cell.

Conclusions

Heterotrophic bacterioplankton in seawater appear to be well adapted to the nutrient regime they encounter. Their small cell size and periplasmic proteins may facilitate nutrient uptake, whereas the presence of multiple or multiphasic transport systems impart metabolic flexibility in a fluctuating nutrient environment. Chemoreception and motility are likely to be of fundamental importance for marine bacterioplankton. Little quantitative information exists regarding these behavioral responses of marine bacteria *in situ*, and this needs to be studied. Although this discussion applies mainly to the euphotic zone of the sea, some general principles may apply to the deeper parts of the sea as well. However, we know little about the mechanisms of UDOM production in the deeper parts of the water column, and this is a fundamental question in determining the bacteria-organic matter interactions in the deep sea. Sinking organic particles are generally considered the main source of UDOM, and it is likely that exoenzymes of particle-associated bacteria play a central role in their utilization.

References

Ammerman JW and Azam F (1981) Dissolved cyclic adenosine monophosphate (cAMP) in the sea and uptake of cAMP by marine bacteria. Mar. Ecol. Prog. Ser. 5 : 85-89.

Ammerman JW and Azam F (1982) Uptake of cyclic AMP by natural populations of marine bacteria. Appl. Environ. Microbiol. 43 : 869-876.

Ammerman JW, Fuhrman JA, Hagstrom A, and Azam F (in press) Bacterioplankton growth in seawater. I. Growth kinetics and cellular characteristics in seawater cultures. Mar. Ecol. Prog. Ser.

Azam F, Fenchel T, Field JG, Gray JS, Meyer-Reil LA and Thingstad F (1983) The ecological role of water-column microbes in the sea. Mar. Ecol. Prog. Ser. 10 : 257-263.

Azam F and Ammerman JW (1984) Cycling of organic matter by bacterioplankton in pelagic marine ecosystems: microenvironmental considerations. In : Fasham MJ (ed.) Flows of energy and materials in marine ecosystems. Theory and practice. NATO-ARI, May 1982 (in press)

Azam F and Hodson RE (1981) Multiphasic kinetics for D-glucose uptake by assemblages of natural marine bacteria. Mar. Ecol. Prog. Ser. 6 : 213-222.

Bell WH and Mitchell R (1972) Chemotactic and growth responses of marine bacteria to algal extracellular products. Biol. Bull. 143 : 265-277.

Burney CM, Johnson KM and Sieburth J McN (1981) Diel flux of dissolved carbohydrate in a salt marsh and a simulated estuarine ecosystem. Mar. Biol. 63 : 175-187.

Franko DA and Wetzel RG (1980) Cyclic adenosine -3':5'-monophosphate: production and extracellular release from green and blue-green algae. Physiol. Plant. 49 : 65-67.

Franko DA and Wetzel RG (1981) Dynamics of cellular and extracellular cAMP in Anabaena Flos-Aquae (cyanophyta): Intrinsic culture variability and correlation with metabolic variables. J. Phycol. 17 : 129-134.

Fuhrman JA (1981) Influence of method on the apparent size distribution of bacterioplankton cells: epifluorescence microscopy compared to scanning electron microscopy. Mar. Ecol. Prog. Ser. 5 : 103-106.

Fuhrman JA and Azam F (1980) Bacterioplankton secondary production estimates for coastal waters of British Columbia, Antarctica and California. Appl. Environ. Microbiol. 39 : 1085-1095.

Fuhrman JA and Azam F (1982) Thymidine incorporation as a measure of heterotrophic bacterioplankton production in marine surface waters: Evaluation and field results. Mar. Biol. 66 : 104-120.

Hagström A, Ammerman JW, Henrichs S and Azam F (1984) Bacterioplankton growth in seawater: II. Organic matter utilization during steady-state growth in seawater cultures. Mar. Ecol. Prog. Ser. (in press)

Hagström A, Larsson U, Hörstedt P and Normark S (1979) Frequency of dividing cells, a new approach to the determination of bacterial growth rates in aquatic environments. Appl. Environ. Microbial. 37 : 805-812.

Hollibaugh JT and Azam F (1983) Microbial degradation of dissolved proteins in seawater. Limnol. Oceanogr. 28 : 1104-1106.

Lampert W (1978) Release of dissolved organic carbon by grazing zooplankton. Limnol. Oceanogr 23 : 831-834.

Nissen H, Nissen P and Azam F (1984) Multiphasic uptake of D-glucose by an oligotrophic marine bacterium. Mar. Ecol. Prog. Ser. (in press)

Sieburth JMcN (1968) Observations on planktonic bacteria in Narragansett Bay, Rhode Island: a resume. Misaki. Mar. Biol. Inst. Kyoto Univ. 12 : 49-64.

Smith DF and Higgins HW (1978) An interspecific regulatory control of dissolved organic carbon production by phytoplankton and incorporation by microheterotrphs. In: Loutit MW and Miles JAR (eds.) Microbial Ecology. Springer-Verlag, Berlin.

Stock JB, Rauch B and Roseman S (1977) Periplasmic Space in *Salmonella Typhimurium* and *Escherichia coli*. J. Biol. Chem. 252 : 7850-7861.

Wiebe WJ and Smith DF (1977) Direct measurement of dissolved organic carbon release by phytoplankton and incorporation by microheterotrophs. Mar. Ecol. 42 : 213-223.

Williams PJLeB (1981) Incorporation of microheterotrophic processes into the classical paradigm of the planktonic foodweb. Kieler Meeresforsch. Sanderh. 5 : 1-28.

Wright RT and Burnison BK (1979) Heterotrophic activity measured with radiolabeled organic substrates. In: Costerton JW and Colwell RR (eds.) Native Aquatic Bacteria: Enumeration, activity and Ecology. Am. Soc. for Testing and Materials, STP695 : p 140-155.

Acknowledgements

This work was supported by NSF grant OCE-26458 and a DOE contract, DE-AT03-82-ER60031.

PHYTOPLANKTON SOLVED THE ARSENATE-PHOSPHATE PROBLEM

A.A. BENSON

Marine Biology Research Division, Scripps Institution of Oceanography, La Jolla, California 92093, USA.

Phosphate is so effectively consumed by photosynthetic activity in surface waters of much of the ocean that its concentration diminishes to that of the ubiquitous arsenate, 10^{-8} M. Since both ions are transported into algae by the same carrier system, all algae are exposed to the hazard of arsenic poisoning. At some early stage of evolution, however, aquatic plants developed an efficient mechanism for detoxifying and excreting arsenic absorbed in their quest for phosphate. Only one such process appears to have evolved. It is so rapid that only very low arsenic levels accumulate in most algae.

The major arsenical metabolites of all phytoplankton were identified by Edmonds and Francesconi (1981) as 5-dimethylarsenosoribosides of glycerol and glycerol sulfate. These strange arsenicals accumulate in all aquatic plants (Nissen and Benson ,1982) in widely varying amounts. A third major arsenical, the arsenolipid isolated radiochromatographically by Cooney et al. (1978), was characterized as a glycerolphosphatide by virtue of its chemical and enzymatic reactivities. Its arsenical moiety, first considered to be a trimethylarsonium derivative, was recognized as a 5-dimethylarsenosoriboside of phosphatidyl glycerol (Edmonds and Francesconi, 1981; Benson and Maruo, 1958), Fig. 1. The

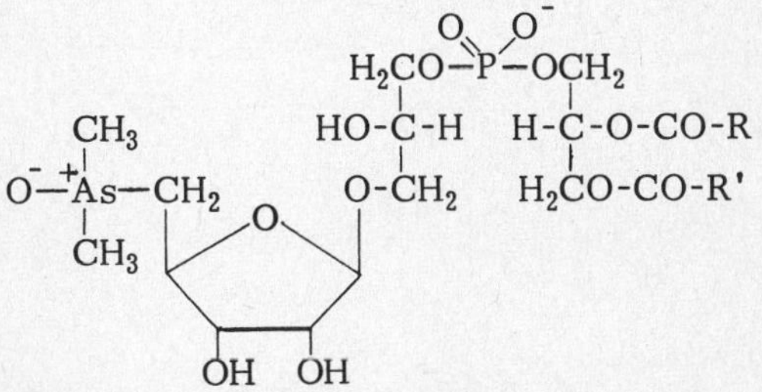

Fig. 1   Fig. 1 The arsenophospholipid, 3-O-5'-dimethylarsenoso- 5'-deoxyribosylphosphatidylglycerol

arsenophospholipid occurs in all aquatic plants in even more variable amounts than do its water-soluble derivatives. Only _Dunaliella_ species and the representatives of the Chlorococcales have arsenolipids besides the arsenoriboside of phosphatidylglycerol. The structure of the second major (50%) arsenolipid of _Dunaliella_ appears to be very different (Cooney, 1981, Wrench and Addison, 1981).

Arsenic Turnover Rates in Phytoplankton

Arsenic levels in most oceanic algal samples have been remarkably low, 2-80 ppm dry wt. In some cases such as _Sargassum_ species, higher arsenic levels have been reported. There appears to be an inverse relationship between environmental phosphate concentrations and the accumulation of arsenic by algae. Such accumulation is related to arsenate uptake and arsenic excretion rates for the organisms. Although no direct measurements of arsenate uptake and arsenic excretion rates have been made, the excellent data of Sanders (1983) and Sanders and Windom, 1980, for arsenate uptake by phytoplankton allow estimation of arsenic turnover under phosphate-sufficient nutrient conditions. Using Sanders' data for arsenate uptake by _Skeletonema costatum_ and the concentration of arsenic in such algae, a turnover of 8 percent per hour is derived. More reliable turnover measurements for algae under a range of nutrient stress conditions are desirable.

The Precursor of Arsenobetaine

The classic work of Edmonds et al., 1977, 1981 and Cannon et al. 1981, reported isolation and identification of arsenobetaine, $(CH_3)_3\overset{+}{As}-CH_2-COO^-$, as the arsenical responsible for the 29 ppm arsenic found in commercial Western Australian rock lobster tail muscle and probably in other organisms (Vaskovsky et al. 1972). Their discovery allayed concern for arsenic toxicity hazards since Welch (1938, 1942) and coworkers, forty years earlier, had studied the metabolism of arsenobetaine and arsenocholine with laboratory rats. At levels of 1% of the diet, no deleterious effects were noted after one week. Other studies have indicated that natural dietary arsenic is readily excreted by the mammalian kidney.

Production of arsenocholine in a marine food chain requires synthetic production of the trimethylarsonium group. The dimethyarsenosoribosides being trialkylarsineoxides are not obvious precursors of trimethylarsonium compounds. Rather, they are oxidation products. Since the identification of arsenobetaine in 1977, no precursors have been recognized.

Isolation of a Novel Arsenical Precursor of Arsenobetaine

In one diatom, Chaetoceros gracilis (Thomas, 1966), of the many species studied to date, there is formed a novel water-soluble arsenical in addition to those previously recognized in all aquatic plants. Whether it is actually the product of symbiotic bacteria is not firmly established. The diatom extracts however, were prepared from radiolabeled cells centrifuged from their media and washed with distilled water. No differences were observed in extracts of unwashed cells.

The novel compound, "D," was neutral at pH 6 and pH 2. It hydrolyzed in dilute acid (0.1N HCl) at 80° in one hour to yield a cationic product having chromatographic properties similar to those of "B," 5-dimethylarsenosoribosylglycerol. Its electrophoretic mobility suggested molecular size and structure of a riboside. "D," was converted in good yield to "A" the 5-dimethylarsenosoribosylglycerol sulfate by bromine oxidation. Similarly, the cationic hydrolysis product of "D" was oxidized by bromine to yield a product with properties of "B." It is tentatively concluded that "D" is a 5-trimethylarsonium-ribosylglycerolsulfate, Fig. 2. As such it is a conceivable metabolic precursor of arsenobetaine.

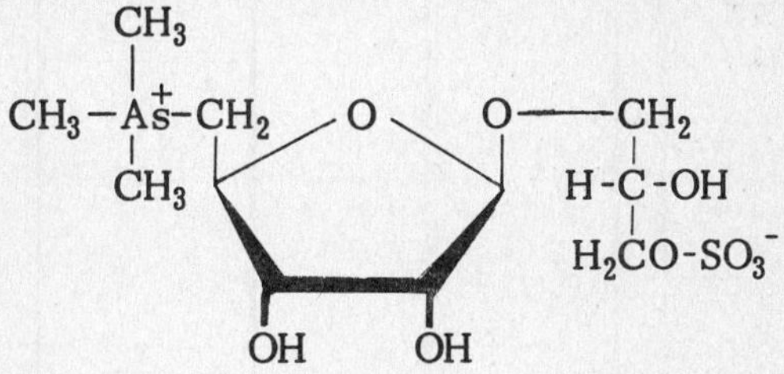

Fig. 2 Proposed Structure of Arsenical Product "D."

Energy Requirements for Arsenic Detoxification

Algal productivity in oligotrophic waters is phosphorus limited. As such conditions lend to comparable rates of arsenate uptake, the productivity is reduced by the energy requirement for arsenic detoxification and excretion. Cacodylate, a seawater component and product of algal arsenate metabolism, is one of the excretion products of radiolabeled phytoplankton species. Its release from membrane-associated arsenophospholipid by oxidation is a plausible process. Extramural bacterial oxidative systems may play a role in its release, Fig. 3.

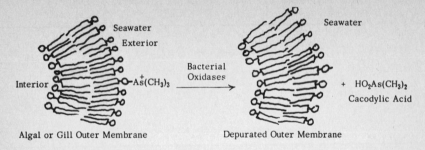

Fig. 3 Arsenic Release from Membrane Lipid Bilayer.

The processes involved in detoxifying arsenate after its absorption by phytoplankton are not firmly established. They appear to be nearly identical in all aquatic plants, suggesting a single evolutionary development. Like phosphates and sulfates, arsenate may be fixed by ADP. Its reduction to the arsonous level has been discussed by Knowles and Benson, 1983. This avid sulfhydryl-binding reagent must be fixed to protein in the chloroplast where (cf. Knowles and Benson, 10983a) it is successively methylated and adenosylated, ultimately producing the 5-dimethylarsenosoribosyl derivatives accumulating in algae.

Energy requirements for the above process include photophosphorylation and reduced nucleotides, NADPH and NADH, produced in the chloroplast. In algae of oligotrophic waters, especially the highly illuminated tropical areas the limited growth and phosphate permit utilization of excess phosphorylating and reducing capabilities in endeavors such as arsenic reduction and methylation. It does not seem likely that arsenic detoxification in oligotrophic waters might reduce productivity significantly. In nutrient-rich waters on the other hand where productivity is light energy-limited, the metabolic energy cost of arsenic detoxification would reduce productivity. The extent of this reduction should approximate the yet unmeasured rate of arsenic processing.

References

Benson AA, and Nissen P 1982. Developments in Plant Biology, $\underline{8}$, 121-124.

Benson AA, and Maruo B 1958. Plant Phospholipids I. Identification of the Phosphatidyl Glycerols. Biochim. et Biophys. Acta, $\underline{27}$, 189-195.

Cannon JR, Edmonds JS et al., 1981. Isolation, crystal structure and synthesis of arsenobetaine, a constituent of the Western Rock Lobster, the Dusky Shark and some samples of human urine. Aust.J.Chem. $\underline{34}$ 787-798.

Cooney RV 1981. The metabolism of arsenic by marine organisms. Doctoral Dissertation, 92 pp. University of California, San Diego.

Cooney RV, Mumma RO et al., 1978. Arsoniumphospholipid in Algae. Proc.Nat.Acad.Sci., 75, 4262-4264.

Edmonds JS, Francesconi KA, et al., 1977. Isolation, crystal structure and synthesis of arsenobetaine, the arsenical constituent of the Western Rock Lobster, Panulirus longipes cygnus George. Tetrahedron Letters, 18 1543-1546.

Edmonds JS, and Francesconi KA 1981. Arsenosugars from the brown kelp (Ecklonia radiata) as intermediates in cycling of arsenic in a marine ecosystem. Nature, 289:602-604.

Knowles FC, and Benson AA 1983. Mode of action of a herbicide. johnson- grass and methanearsonic Acid. Plant Physiology, 71, 235-240.

Knowles FC, and Benson AA 1983. The biochemistry of arsenic. Trends in Biochemical Sciences, 8, 178-180.

Nissen P, and Benson AA 1982. Arsenic metabolism in freshwater and terrestrial plants. Physiologia Plantarum 54, 446-450.

Sanders JG 1983. Role of marine phytoplankton in Determining the Chemical Speciation and biogeochemical cycling of arsenic. Can.J.Fish.Aquatic Sci. 40.

Sanders JG, and Windom HL 1980. The uptake and reduction of arsenic species by marine algae. Estuar. and Coastal Mar. Sci. 10, 555-567 (1980).

Thomas WH 1966. Effects of temperature and illuminance on cell division rates of three species of tropical oceanic phytoplankton. J. Phycol., 2, 17-22.

Vaskovsky VE, Korotchenko OD, et al., 1972. Arsenic in the lipid extracts of marine invertebrates. Comp.Biochem.Physiol. 41B, 777-784.

Welch AD, and Welch MS 1938. Lipotropic action of choline derivatives. Proc. Soc. Exp. Biol.Med. 39 7-9.

Welch AD, and Landau RL 1942. The arsenic analog of choline on a component of lecithin in rats fed arsenocholine chloride. J. Biol.Chem. 144 581-588.

Wrench JJ, and Addison RF 1981. Reduction, methylation and incorporation of arsenic into lipids by the marine phytoplankton Dunaliella tertiolecta. Can.J.Fish.Aquat.Sci. 38, 518-523.

EXCRETION OF ORGANIC CARBON AS FUNCTION OF NUTRIENT STRESS

A. JENSEN

Institute of Marine Biochemistry, University of Trondheim,
N-7034 Trondheim, NTH, Norway

INTRODUCTION

For half a century aquatic biologists have studied and debated how planktonic algae contribute to the pool of dissolved organic matter in culture media and in natural waters. Krogh took up this topic already in 1934 (Krogh, 1934) and the debate seemed to culminate with Sharp's contribution in 1977 immediately followed by Fogg's comments (Sharp, 1977; Fogg, 1977) and by the duel between Aaronson and Sharp in 1978 Aaronson, 1978; Sharp, 1978). Several reviews have covered the development of the field (Fogg, 1971; Hellebust, 1974; Wangersky, 1978; Sharp, 1977), and more recent publications have clarified some of the discrepancies (Nalewajko, 1977; Mague et al., 1980; Larsson and Hagström, 1979, 1982).

There is a general agreement that algal cells will deliver dissolved organic compounds to the surrounding water when the cells die or are disrupted in some way, e.g. through grazing ("sloppy feeding"). The major dispute has been concerned with the release of extracellular material by living cells which is the real excretion or exudation. Sharp (1977) concluded that evidence of excessive excretion of organic material by "healthy" phytoplankton was not good, and he maintained (Sharp, 1978) that undisturbed phytoplankton would only excrete negligible quantities of material in nature. By "healthy" cells he meant exponentially growing cells.

It seems to be generally accepted that exponentially growing cells excrete less material than cells in other growth phases do (see p. ex. Wangersky, 1978; Guillard and Wangersky, 1958; Myklestad and Haug, 1972; Haug and Myklestad, 1976; Myklestad, 1977). However, since phytoplankton populations normally go through several growth phases in addition

to the exponential phase, and excretion can be quite significant in pre
- as well as post - exponential phases, excretion must be expected to
have physiological and ecological significance in many cases. There
is ample evidence to indicate that changes in light, temperature, nutrient levels, salinity and population density may induce significant
excretion of organic compounds by otherwise "healthy" cells.
(Guillard and Wangersky, 1958; Marker, 1965; Soeder and Bolze, 1981).
The release of organic compounds by planktonic algae thus depends on
the physiological state of the cells as well as on many environmental
factors. In addition, good evidence has been provided to show clear
differences between algal groups and species. The diatom *Skeletonema
costatum* does not seem to excrete much organic carbon, even in the late
stationary phase, while several *Chaetoceros* species tend to release
large quantities of extracellular carbohydrate in this growth phase
(Myklestad and Haug, 1972; Myklestad, 1974, 1977). Dinoflagellates
seem to be good excretors in genereal (Marker, 1965; Wangersky, 1978).
Differences in behaviour between species may explain some of the dis-
agreements found in the literature. Sharp (1977) based some of his
conclusions on results obtained with *S. costatum*. Other discrepancies
may have originated in the stress conditions unintentionally introduced
in the treatment of the algae. Among the stress factors which may influence excretion, nutrient deficiency is probably the most important.

EXCRETION AND NUTRIENTS

Nutrient levels and kinetics as well as excretion and release rates
are not easily studied in nature. The more conclusive information on
phytoplankton excretion has therefore been provided by studies of algal
cultures under defined conditions in the laboratory.

A wealth of organic compounds that may have come from algal sources
has been found in lake and seawater (for reviews see Duursma, 1965;
Collier, 1970; Wangersky, 1978), but only a limited numbers of compounds
are likely to be involved in excretion. Among these are glycollate
(Hellebust, 1965; Watt, 1969), glycerol (Graigie *et al*., 1966), amino
acids and peptides (Fogg and Westlake, 1955; Hellebust, 1965; Aaronson
et al., 1971; Walsby and Fogg, 1975) vitamin B_{12}, thiamin and biotin
(Carlucci and Bowes, 1970; Aaronson *et al*. 1971, and above all various
types of polysaccharides (Lewin, 1956; Marker, 1965; Hoyt, 1970; Aaronson, 1971; Huntsman, 1972; Myklestad and Haug, 1972; Myklestad *et al*.,
1972, Myklestad, 1974, 1977). Excretion of toxins by dinoflagellates

are well known, and coloured products (Cristofalo et al., 1962) as well as surface-active compounds (Wilson and Collier, 1972) are among the more exotic exudates of microscopic algae.

Certain organic phosphorus compounds are also involved in some form of rapid excretion and uptake by phytoplankton (Lean and Nalewajko, 1976). In addition several enzymes such as a β-glucosidase (Mayer, 1976) as well as lipids (Billmire and Aaronson, 1976) must be regarded as real excretion compounds having important functions in the metabolism and reproduction of algae.

Compounds possibly involved in massive excretion which may play a significant role in the energy flow in lakes and oceans are above all the carbohydrates and to some extent glycollate, amino acids and peptides. The concentration of certain carbohydrates, especially polysaccharides, found in culture media may be quite impressive. Myklestad and Haug (1972) detected 40 mg polysaccharide per liter while Moore and Tischer (1964) as well as Marker (1965) reported values corresponding to 500 mg/l. The highest concentrations were always found in old cultures in some stress situation. Very often old cultures (and algal populations in general) are not only old but also starved, and it is difficult to separate the influence of these two factors on the excretion process.

In the 1970's Myklestad and Haug started a series of investigations on the excretion of carbohydrates by marine diatoms and paid special attention to the influence of the major nutrient salts on this process. Much of our knowledge of the influence of nitrate and phosphate on excretion of carbohydrates by marine phytoplankton originate in these studies which will be treated in some detail in the following. During their investigation of the effect of nutrient composition and concentration in the growth medium on the chemical composition of the diatom *Chaetoceros affinis* Myklestad and Haug (1972) observed that the main production of extracellular polysaccharide followed a period of rapid increase in the cellular content of storage carbohydrates. The latter started when the medium became nitrate depleted and the cells were in the stationary phase. The release of extracellular polysaccharide was enhanced by a high proportion of nitrate to phosphate in the medium. (Fig. 1).

Fig. 1. Release of extracellular polysaccharide by the diatom *Chaetoceros affinis*, according to Myklestad and Haug (1972).

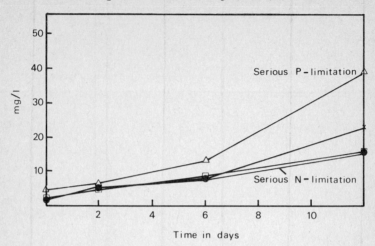

The concentration of extracellular polysaccharide in the medium reached 40 mg/l in these experiments, and the authors noticed that the amount of extracellular polysaccharide produced in a 12-day old culture was three times higher than the cellular glucan production. In a further study of the chemical composition of the extracellular polysaccharide Myklestad *et al.* (1972) established that the content of rhamnose and fucose in the carbohydrate clearly distinguished this polymer from the cellular storage polysaccharides. Leakage from dead or dying cells could therefore not explain the presence of the extracellular polysaccharide in the medium, and the authors concluded that it was released from healthy cells in the stationary growth phase. This conclusion was further substantiated in a study by Haug and Myklestad (1976).

Myklestad (1974) included 8 diatoms in his further studies and found marked differences in the production of cellular glucan and extracellular polysaccharides between the species. In parallel experiments the release of extracellular polysaccharide amounted to some 25 mg/l for two of the *Chaetoceros* species (*C. curvisetus* and *C. affinis*), while negligible quantities (1-2 mg/l) were excreted by *Thalassiosira gravida* and *S. costatum* (see Table I).

TABLE I. Myklestad (1974).

Content of cellular carbohydrate and extracellular polysaccharide in 12-days cultures, mg/l.

	Cellular carbohydr.	Extracellular polysaccharide	Extracell.polys./ cell. carbohydr.
C. affinis	46	23	0.50
C. debilis	19.5	6	0.31
C. curvisetus	22	27.5	1.25
C. socialis	17.5	5	0.29
T. gravida	27	2	0.07
S. costatum	80	1	0.01
T. fluviatilis	115	4	0.03

The ratio of extracellular polysaccharide to cellular carbohydrate varied even more between these 12-day old cultures, which were grown under identical conditions, involving N-limitation after the 6th day The ratio was well below 0.1 for S. costatum and the two Thalassiosira species and between 0.29 and 1.25 for the Chaetoceros members.

A detailed investigation of the influence of the N/P ratio in the growth medium on the production of cellular and extracellular carbohydrates was then carried out on Chaetoceros affinis and S. costatum (Myklestad, 1977). The N/P ratio varied in six steps from 0.4 to 100, and the production of extracellular polysaccharide was studied together with growth rate, assimilation ratio of nitrogen to phosphorus, production of cellular carbohydrate and protein, cellular phosphorus and cell size during the exponential as well as the stationary phase.

A decrease in the growth rate from 1.7 to 1.1 divisions per day took place when the nitrate content in the medium was depleted. Rapidly growing cells of both species produced relatively little carbohydrate while slow growing cells produced more. When the N/P ratio was 100 and the nitrate in the medium more or less used up, C. affinis released extracellular polysaccharide in quantities corresponding to the content of total carbohydrate in the cells, while the release from S. costatum remained small. In a special experiment C. affinis was started in a medium with N/P ratio of 109 (104 µM nitrate and 0.95 µM phosphate). The culture was supplemented four times with nitrate during

22 days to keep the concentration at 100 µM, while no more phosphate was added. Exponential growth, correspondingly to 1.3 divisions per day lasted for 6 days (Fig. 2) while all the phosphate had been taken up by the 4th day.

Fig. 2. Growth of *Chaetoceros affinis* under phosphate limitation. Cell density (-●-, cells/ml x 10^3); cellular carbohydrate (-Δ- mg/l); cellular protein (-○-, mg/l); extracellular polysaccharide (-▲-, mg/l). Myklestad (1977).

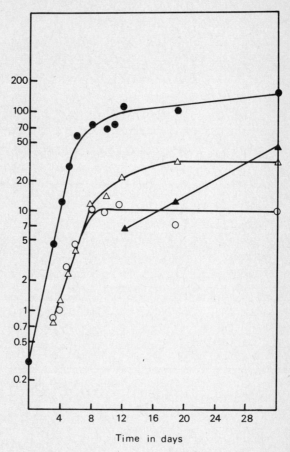

Net protein synthesis stopped after 8 days and the production of cellular carbohydrate continued for another 16 days. The cells kept up their division at a slow rate also after the halt of the protein production. The most striking observation, however, was the fact that the release of extracellular polysaccharide continued at high rate to the end of the experiment and long after the production of cellular carbohydrate had ceased.

The concentration of extracellular polysaccharide in the medium of this *C. affinis* culture increased from 6.5 to 43 mg/l between the 12th and the 32nd day. During the last 13 days of the experiment this was practically the only photosynthate produced by the alga.

Ignatiades and Fogg (1973) also studied the influence of nutrients (N, P and Si together) on the release of organic matter by phytoplankton. These authors were also applying "f" media in dilution series, from "f" to "f"/60 and found that excretion of photosynthate increased from 1% to 65% of photosynthetically fixed carbon as the nutrients were diluted (constant N: P: Si); the main excretion took place after the exponential phase. Since the authors worked with *S. costatum* rather small quantities of material were released. In cultures with $5 \cdot 10^5$ cells/ml which had been cultivated at the lowest nutrient levels (maximum excretion) a little more than 1 g C $m^{-3} h^{-1}$ was produced. Although these experiments demonstrate that nutrient stress favours production of extracellular compounds by planktonic algae, the choice of the alga and small quantities excreted as a consequense of this will not convince the ecologist that algal excretion is important in the energy flow in the sea.

Myklestad and Haug kept reasonable track in their experiments with the concentration of nutrient salts in the medium as the cultures developed, and they varied the nutrients independently. Excretion could therefore to some extent be correlated with nutrient levels in the medium. In general, studies on excretion do not include determination of the concentration of nutrients in the medium throughout the experiments. The frequently observed phenomenon of increased release of organic matter in the stationary phase can therefore only be taken as an indication of a causal relationship with nutrient stress in these cases.

A recent exception is formed by the work of Soeder and Bolze (1981) on the effects of sulphate deficiency on the release of dissolved organic matter in cultures of *Scenedesmus obliquus*. Sulphate deficiency definitely stimulated this release.

The field studies involving natural phytoplankton can also provide only indications and little solid information on the relationship between nutrient levels and algal excretion. In most cases excessive excretion is found at the end of or after the algal bloom. Very often grazing on the algae and consumption of the excretion products have taken place during the experiments, and cell breakage together with other experimental weaknesses occured, in addition to the difficulties caused by uneven distribution of organisms and nutrients. All this, together with the fact that nutrient concentrations as well as phytoplankton species composition are often not determined, make data collected in field experiments and in nature more or less useless in the present context. Many of the problems involved have been pointed out by Berman and Holm-Hansen (1974); Sharp (1977) and Wangersky (1978) among others.

At this point credit should be given to several recent workers, especially Wiebe and Smith (1977), Jüttner and Matuschek (1978), Lancelot (1979) and Larsson and Hagström (1979, 1982) who have developed new and useful methods and provided more evidence to demonstrate that excretion of organic matter by natural phytoplankton is important under special circumstances. These contributions do, however, not shed new light on the role of nutrient stress on the release processes, mainly because they were not specially designed for this aspect.

CONCLUSION

We know that excretion of organic compounds by planktonic algae differ between species in regard both to the nature of the compounds and the quantities released. Nutrient levels and composition undoubtedly have strong influence on the release processes. It is quite clear that some algal species will excrete most of their photosynthate to the surrounding water when the cells have plenty of light, are nutrient limited (N and P) and grow slowly. The quantities released under such circumstances may be quite significant. The dominating component of the excreted material will very likely be carbohydrates. We do not know how widespread this character is among commonly occurring microscopic algae since only a limited number of species have been properly investigated.

We know that N- and P-limited phytoplankton occur in lakes, fjords and coastal waters in temperate and boral zones. It is therefore reason to believe that excretion of organic matter may be quite significant in such waters, provided the species composition of the phytoplankton is right and sufficient light energy is available to drive the process.

If we think we need to know how much organic matter is provided through phytoplankton excretion in natural water we must establish rates for the algal species involved as function of the physiological state of the algae, the nutrient levels in the water and other environmental factors such as temperature, light, turbulence, etc., and the way these vary.

The relationship to nutrient levels and composition is probably best studied in chemostats in the laboratory or even better by the cage culture technique (Jensen *et al.*, 1972), which is ideal for excretion studies, provided the cages are operated in the continuous reservoir mode which secures constant external nutrient levels througout the experiments.

It may strike the reader that the amount of information needed to evaluate the production in natural waters may become unsurmountable if we have to build up our estimates from detailed knowledge of all the processes involved. We do not have to know how every screw is tightened and how much energy this requires in order to establish the production of cars by the automobile industry. Maybe we should look for integrating parameters, and not so much for details in our efforts to quantify aquatic production. Models based on input-output parameters are easier to operate and may be readily tested, in contrast to mechanistic models which will remain oversimplifications, maybe forever.

LITERATURE REFERENCES

Aaronson, S. 1971: The synthesis of Extracellular Macromolecules and Membranes by a Population of the Phytoflagellate *Ochromonas danica*. Limnol. Oceanogr. 16, 1-9.

Aaronson, S. 1978: Excretion of Organic Matter by Phytoplankton *in vitro* Limnol. Oceanogr. 23, 838.

Aaronson, S. DeAngelis, B., Frank, O. and Baker, H. 1971: Secretion of Vitamins and Amino Acids into the Environment by *Ochromonas danica*. J. Phycol. 7, 215-218.

Berman, T. and Holm-Hansen, O. 1974: Release of Photoassimilation Carbon as dissolved Organic Matter by Marine Phytoplankton. Marine Biol. 28, 305-310.

Billmire, E. and Aaronson, S. 1976: The Secretion of Lipids by the Freshwater Phytoflagellate *Ochromonas danica*. Limnol. Oceanogr. 21, 138-140.

Carlucci, A.F. and Bowes, P.M. 1970: Production of Vitamin B_{12}, Thiamine and Biotin by Phytoplankton. J. Phycol. 6, 351-357.

Collier, A. 1970: Oceans and Coastal Waters as Lifesupporting Environments. O. Kinne (ed.) Marine Ecology, Vol 1. Environmental Factors,' Part 1, Wiley, London, pp. 1-93.

Craigie, J.S., McLachlan, J., Majak, W., Ackman, R.G. and Tocher, C.S. 1966: Photosynthesis in Algae. II. Green Algae with special Reference to *Dunaliella* spp. and *Tetracelmis* spp. Can. J. Bot. 44, 1247-1254.

Christofalo, V.J., Kornreich, L.D. and Ronkin, R.R.: *Chlamydomonas*: Colored Excretion products. Science, 1938, 809-810.

Duursma, E.K. 1965: The Dissolved Organic Constituents of Sea Water In P.J. Riley and Skirrow, G. (eds.). Chemical Oceanogr. Vol. 1, Acad. Press. N.Y. pp. 433-475.

Fogg, G.E. 1971: Extracellular Products of Algae in Fresh Water. Arch. Hydrobiol. Beih. Ergeb. Limnol. 5, 1-25.

Fogg, G.E. 1977: Excretion of Organic Matter by Phytoplankton. Limnol. Oceanogr. 22, 576-577.

Fogg, G.E. and Westlake, D.E. 1955: The Importance of Extracellular Products of Algae in Fresh Waters. Verh. int. Ver. Limnol. 12, 219-231.

Guillard, R.R.L. and Wangersky, P.J. 1958: The Production of Extracellular Carbohydrates by Some Marine Flagellates. Limnol. Oceanogr. 3, 449-454.

Haug, A. and Myklestad, S. 1976: Polysaccharides of Marine Diatoms with Special Reference to *Chaetoceros* Species. Mar. Biol. 34, 217-222.

Hellebust, J.A. 1965: Excretion of Some Organic Compounds by Marine Phytoplankton, Limnol. Oceanogr. 10, 192-206.

Hellebust, J. A. 1974: Extracellular Products. In W.D.P. Stewart (ed.). Algal Physiology and Biochemistry. Bot. Monogr. Vol. 10, Univ. Calif. Press., Berkeley, pp. 838-863.

Hoyt, J.W. 1970: High Molecular Weight Substances in the Sea. Mar. Biol. 7, 93-99.

Huntsman, S.A. 1972: Organic Excretion by *Dunaliella tertiolecta*. J. Phycol. 8, 59-63.

Ignatiades, L. and Fogg, G.E. 1973: Studies on the Factors Affecting the Release of Organic Matter by *Skeletonema costatum* (Grew.) Cleve in Cultures. J. Mar. Biol. Ass. U.K. 53, 937-956.

Jensen, A., Rystad, B. and Skoglund, L. 1972: The use of Dialysis Culture in Phytoplankton Studies. J.Exp.Mar. Biol. 8, 241-248.

Juttner, F. and Matuschek, T. 1978: The Release of Low Molecular Weight Compounds by the Phytoplankton in an Eutrophic Lake. Water. Res. 12, 251-257.

Krogh, A. 1934: Conditions of Life in the Oceans. Ecol. Monogr. 4, 421-429.

Lancelot, C. 1979: Gross Excretion Rates of Natural Phytoplankton and Heterotrophic Uptake of Excreted Products in the Southern North Sea, as Determined by Short-Time Kinetics. Mar. Ecol. 1, 179-186.

Larsson, U. and Hagström, A. 1979: Phytoplankton Exudate Release as an Energy Source for the Growth of Pelagic Bacteria. Mar. Biol. 52, 199-206.

Larsson, U. and Hagström, A. 1982: Fractionated Phytoplankton Primary Production, Exudate Release and Bacterial Production in a Baltic Eutrophication Gradient. Mar. Biol. 67, 57-71.

Lean, D.R.S. and Nalewajko, C. 1976: Phosphate Exchange and Organic Phosphorus Excretion by Freshwater algae. J. Fish. Res. Bd. Canada, 33, 1312-1323.

Lewin, R.A. 1956: Extracellular Polysaccharide of Green Algae. Can. J. Microbiol. 2, 665-672.

Mague, T. H., Friberg, E., Hughes, D.J. and Morris, I. 1980: Extracellular Release of Carbon by Marine Phytoplankton: A Physiological Approach. Limnol. Oceanogr. 25, 262-279.

Marker, A.F.H. 1965: Extracellular Carbohydrate Liberation in the Flagellates *Isochrysis galbana* and *Prymnesium parvum*. J. Mar. Biol. Ass. U.K. 45, 755-772.

Meyer, D.H. 1976: Secretion of β-Glucosidase by *Ochromonas danica*. Arch. Microbiol. 109, 263-270.

Moore, G.B. and Tischer, R.G. 1964: Extracellular Polysaccharides of Alga : Effects on Life-Support Systems. Science, 145, 586-587.

Myklestad, S. 1974: Production of Carbohydrates by Marine Planktonic Diatoms. I. Comparison of Nine Different Species in culture. J. Exp. Mar. Biol. Ecol. 15, 261-274.

Myklestad, S. 1977: Production of Carbohydrates by Marine Planktonic Diatoms. II. Influence of the N/P ratio in the Growth Medium on the Assimilation Ratio, Growth Rate, and Production of Cellular and Extracellular Carbohydrates by *Chaetoceros affinis* var. *Willei* (Gran.) Hustedt and *Skeletonema costatum* (Grev.) Cleve. J. Exp. Mar, Biol.Ecol. 29, 161-179.

Myklestad, E. and Haug, A. 1972: Production of Carbohydrates by the Marine Diatom *Chaetoceros affinis* var. *Willei* (Gran.) Hustedt. I. Effect of the Concentration of Nutrients in the Culture Medium. J. exp. mar. Biol Ecol. 9, 125-136.

Myklestad, S., Haug, A. and Larsen B. 1972: Production of Carbohydrates by the Marine Diatom *Chaetoceros affinis* var. *Willei* (Gran.) Hustedt. II. Preliminary Investigation of the Extracellular Polysaccharide. J. exp. mar. Biol. Ecol. 9, 137-144.

Nalewajko, C. 1977: Extracellular Release in Freshwater Algae as a Source of Carbon for Heterotrophs. In J. Cairns Jr. (ed.) Aquatic Microbial Communities. Garland Publ. Inc. London, pp. 589-642.

Sharp, J.H. 1977: Excretion of Organic Matter by Marine Phytoplankton: Do Healthy Cells Do It? Limnol. Oceanogr. 22, 381-399.

Sharp, J.H. 1978: Reply to Comments by S. Aaronson. Limnol. Oceanogr. 23, 839-840.

Soeder, C.J. and Bolze, A. 1981: Sulphate Deficiency Stimulates Release of Dissolved Organic Matter in Synchronous Cultures of *Scenedesmus obliquus*. Physiol. Plantarum 52, 233-238.

Walsby, A.E. and Fogg, G.E. 1975: The Extracellular Products of *Anabaena cylindrica*. III. Excretion and Uptake of Fixed Nitrogen. Br. Phycol. J. 10, 339-345.

Wangersky, P.J. 1978: Production of Dissolved Organic Matter. In O. Kinne (ed.) Marine Ecology, Vol. 4, Wiley, London, 115-200.

Watt, W.D. 1969: Extracellular Release of Organic Matter from two Freshwater Diatoms. Ann. Bot. 33, 427-437.

Wiebe, W.J. and Smith, D.F. 1977: Direct Measurement of Dissolved Organic Carbon Release by Phytoplankton and Incorporation by Microheterotrophs. Mar. Biol. 42, 213-225.

Wilson, W.B. and Collier, A. 1972: The Production of Surface-Active Material by Marine Phytoplankton Cultures, J. Mar. Res. 30, 15-26.

SEASONAL CHANGES IN PRIMARY PRODUCTION AND PHOTOADAPTATION BY THE REEF-BUILDING CORAL ACROPORA GRANULOSA, ON THE GREAT BARRIER REEF

B.E. CHALKER, T.COX and W.C.DUNLAP

Australian Institute of Marine Science, P.M.B.n°3,
Townsville M.C., Queensland 4810, Australia

SUMMARY

Using previously developed equations, instantaneous rates of in situ photosynthesis were estimated for Acropora granulosa in December and June at Davies Reef, Great Barrier Reef, Australia (147°38'E, 18°51'S). Integration yielded estimated daily rates. Diel P/R ratios are high from the surface to 35 m (5.8% of daily surface light) and decline rapidly thereafter. Diel P/R ratios are greatest at a depth of 25 m (12% of daily surface light). During December the diel compensation depth is 60 m (1% of daily surface light); during June, it is 45 m (1.6% of daily surface summer light). Thus, it can be expected that Acropora granulosa will do well metabolically from the surface to a depth transmitting 5.8% of surface light; will do poorly from that depth down to a depth transmitting 1% of surface light; and is unlikely to survive at a depth below the 1% light level. These metabolic expectations coincide with observed abundances from 6 to 51 m.

INTRODUCTION

Reef building corals contain large populations of the endosymbiotic dinoflagellate Gymnodinium microadriaticum (Freudenthal). The presence of the alga has profound physiological consequences. A significant fraction of photosynthetically fixed carbon is released from the algae and metabolised by the animal tissues (Muscatine and Cernichiari, 1969; Smith, Muscatine and Lewis, 1969; Muscatine, Pool and Cernichiari, 1972; Muscatine, 1980). In addition, algal photosynthesis stimulates the rate of coral calcification (Goreau, 1959; Goreau and Goreau, 1959; Vandermeulen, Davis and Muscatine, 1972; Vandermeulen and Muscatine, 1974; Chalker and Taylor, 1975).

The relationship between light intensity (I) and photosynthesis (P) can be examined experimentally by the construction of light-saturation (P-I) curves. All coral P-I curves have a similar shape. Initially, photosynthesis is directly proportional to light intensity. Thereafter, the curve rapidly approaches a horizontal asymptote (P_m) which is variously described as the photosynthetic capacity, photosynthetic maximum, or assimilation number. It is generally assumed that the initial slope of the curve (α) is a function of the light reactions in photosynthesis, and that the slope of the curve declines when the rate of cellular carbon metabolism becomes limiting (Steemann Nielsen and Jørgensen, 1968). The irradiance at which the initial slope of the curve intercepts the horizontal asymptote is defined as I_k (Talling, 1957). Photosynthesis as represented in P-I curves is usually gross photosynthesis (P^g) which is defined as the sum of net photosynthesis (P^n) and the dark respiration rate (R).

Light saturation curves for photosynthesis can be described by the equations (adapted from MacCaull and Platt, 1977):

$$P^n = P_m^g \tanh (I/I_k) + R \tag{1}$$

and

$$P^n = P_m^g \tanh (\alpha I / P_m^g) + R \tag{2}$$

Experimentally determined values for R are always negative to indicate that oxygen is consumed during respiration.

In suitable habitats, reef-building corals are abundant at all depths within the photic zone or approximately from 100% to 1% of the surface light intensity. It is not surprising that photosynthetic photoadaptations have been observed in comparisons of P-I curves for corals growing at different depths (Kawaguti, 1937; Wethey and Porter, 1976a,b; Davies, 1977,1980), in light and shade environments at the same depth (Falkowski and Dubinsky, 1981), or both (Zvalinskii et al., 1980). In general corals growing in high light environments have greater values for P_m^g (expressed as oxygen produced per hour per mg chlorophyll-a, I_k, and R than do corals growing in low light environments. Conversely corals growing in low light environments have greater values for α and are thus potentially more effective photosynthetically in dim light than are their conspecifics from high light environments.

The relationships between the seasonal average values for these photokinetic parameters and light intensity (photosynthetic photon flux density, PPFD, 400-700 nm), can be described by the general equation,

$$\ln X_T = (\text{slope}) \ln T + \ln X_1 \tag{3}$$

where X is a photokinetic parameter, T is the function of PPFD transmitted to the collection depth, and $\ln X_1$ is the y-intercept (Chalker et al., 1983). The differences between summer and winter values for the photokinetic parameters are strictly a function of light intensity. Thus, equation (1) is valid for both summer and winter if T is redefined to the fraction of light reaching each depth relative to light just below the surface during a cloudless summer solstice (Chalker and Dunlap, 1983a). The possibility of hysteresis in the approach to these seasonal extremes, such as that observed for phytoplankton (Steemann Nielsen and Hansen, 1961), has not yet been tested.

This communication uses these mathematical descriptions of coral photoadaptation to simulate light saturation curves for the reef-building coral Acropora granulosa growing at various depths on Davies Reef, Great Barrier Reef during December and June. Diel curves are also constructed and integrated to estimate total diel gross photosynthesis and respiration. These values are used to predict the nutritional status of A. granulosa at different depths on the reef, and predicted nutritional status is compared with observed abundance.

METHODS

Instantaneous in situ rates of photosynthesis by A. granulosa growing at different depths on Davies Reef during the summer (December) and winter (June) were estimated using the model of Chalker and Dunlap (1983b). Light saturation curves for photosynthesis were simulated by equation (1). P_m^g, I_k, and R were calculated for the light intensities corresponding to each depth by the formulae :

$$P_{mT}^g = P_{m1}^g \exp(c \ln T), \tag{4}$$

$$I_{kT} = I_{k1} \exp(a \ln T), \tag{5}$$

and

$$(-R_T) \simeq (-R_1) \exp(d \ln T). \tag{6}$$

The logarithmic forms of equations (4-6) are developed in Chalker et al., (1983). The values of the slopes and y-intercepts were calculated from data presented by Chalker and Dunlap (1983a).

Daily changes in solar photosynthetic photon flux density 400-700 nm (PPFD) were simulated by the equation

$$I = I_m \sin(\pi H/D), \tag{7}$$

where I_m is the maximum irradiance at local solar noon, H is the hour of the day, and D is the length of the day from civil twilight in the morning until civil twilight in the evening. Values for D were obtained from the Nautical Almanac (Her Majesty's Nautical Almanac Office, 1980). Values for I_m at each depth were obtained by multiplying surface PPFD at local solar noon by the measured light transmittance of the water at each depth. Estimated *in situ* rates of net photosynthesis were calculated by substituting equations (4-7) into equation (1), to yield

$$P_T^g = P_{ml}^g \exp(c \ln T) \tanh(I_{mT} \sin(\pi H/D)/(I_{kl} \exp(a \ln T))) - R_1 \exp(d \ln T) \quad (8)$$

Light saturation curves for gross photosynthesis were simulated by the equation,

$$P^g = P_m^g \tanh(I/I_k). \quad (9)$$

A single equation for estimating instantaneous rates of gross photosynthesis was obtained by combining equations (4,5,7 and 9),

$$P_T^g = P_{ml}^g \exp(c \ln T) \tanh(I_{mT} \sin(\pi H/D)/(I_{kl} \exp(d \ln T))). \quad (10)$$

Estimated diel *in situ* rates for primary production were calculated by integrating equation (10) for the length of the day using a Simpson's rule approximation (Hamming, 1971). Estimated *in situ* rates of respiration were calculated by multiplying hourly rates, from equation (6), by twenty-four.

RESULTS AND DISCUSSION

Using equations (1,4,5, and 6) and parameters calculated from the data of Chalker and Dunlap (1983a), light saturation curves for photosynthesis were calculated for <u>A. granulosa</u> growing at depths from 1 to 50 m on Davies Reef in December (figure 1) and June (figure 2). In each figure, P_m^g (normalized per mg protein) and α both increase with increasing depth; R is constant; and I_k declines. The transition from relatively high irradiance in summer to relatively low irradiance in winter produces analogous changes in the values of the photokinetic parameters: P_m^g and α both increase, R is constant, and I_k declines.

In general, these observations are consistent with photosynthetic photoadaptations which are commonly reported. The only exception is the observation that R is constant for A. granulosa at all depths and seasons. For Acropora spp. (Chalker et al., 1983) and for other species (Kawaguti, 1937; Davies 1977, 1980) respiration declines with decreasing irradiance.

Estimated in situ rates of instantaneous net photosynthesis were calculated from equation (8). Results for December and June are illustrated in figure 3 and 4, respectively. There is a newly homeostatic carbon flux throughout the day from the surface to a depth of 30 m. Below 40 m, due to reduced PPFD, net photosynthesis never closely approximates its potential maximum. Seasonally, summer values are considerably higher at any given hour than are winter values. In addition, the decline in instantaneous net photosynthesis below 30 m, is far more dramatic in winter than in summer.

The shapes of the curves in figures 3 and 4 are typical of those which have been previously observed for corals by means of in situ respirometers (Wells, 1977). They are also similar to curves produced by the models of Vollenweider (1965) when the possibility of photoinhibition is not included.

Diel gross photosynthesis for December and June as calculated from the integrals of equation (10) are illustrated in figure (5). During the summer, diel P/R ratios increase with increasing depth from the surface to 25 m (12% of daily surface light), decline slowly to 35 or 40 m (5.8 - 4.1% of daily surface light), and are projected to decline rapidly thereafter. During December, the diel compensation depth is 60 m (1% of daily surface light).

During June, diel P/R ratios increase from the surface to a depth of 15-20 m (13 - 9.1% of daily summer surface light), declines slowly to 30 m (4.5% of daily summer surface light), and declines rapidly thereafter. During June, the compensation depth is 45 m (1.6% of daily summer surface light).

Thus, it can be expected that Acropora granulosa will do well metabolically from the surface to a depth transmitting approximately 5% of the irradiance at the surface of the summer; will do poorly from that depth down to a depth transmitting 1% of summer surface light; and is unlikely to survive below the 1% light level. These metabolic expectations coincide with subjectively observed abundances from A. granulosa from depths of 6-51 m on Davies Reef.

It is universally recognized that most reef-building corals possess considerable heterotrophic potential (Yonge, 1973); but the relative importance of heterotrophy to total coral metabolism, has never been quantified in the field. Our calculations and field observations of Acropora granulosa suggest that any heterotrophy which might occur does not result in an extension of the range of Acropora granulosa below the depths that are predicted based upon known rates of respiration and photosynthesis. These studies emphasise the predictive value of quantitative metabolic studies and the importance of light as an ecological parameter affecting reef-building corals and their symbionts.

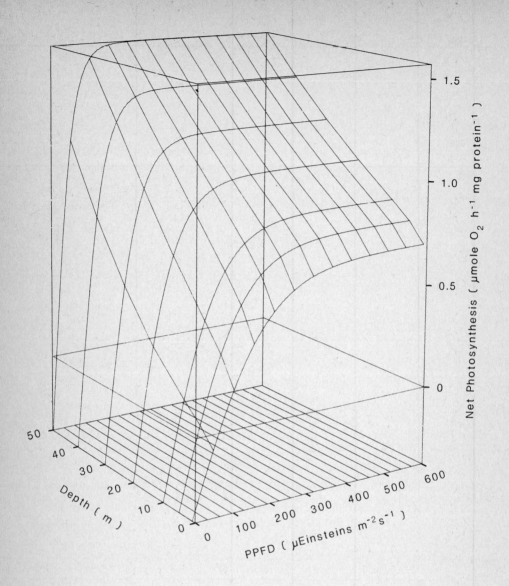

Figure 1. Calculated light-saturation curves for <u>Acropora granulosa</u> at depths of 1, 5, 10, 20, 30, 40, and 50 m on Davies Reef during December.

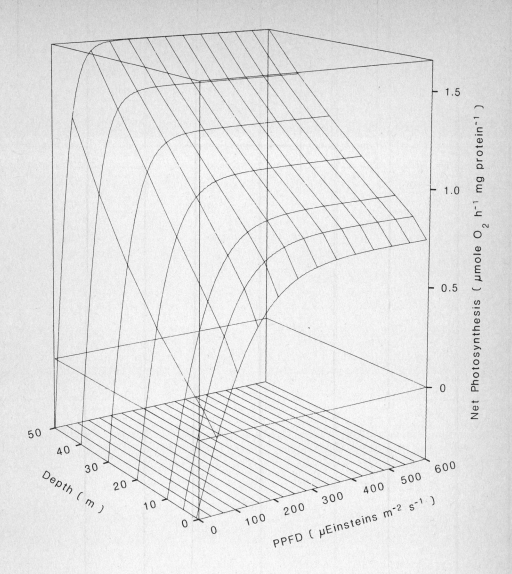

Figure 2. Calculated light-saturation curves for <u>Acropora granulosa</u> at depths of 1, 5, 10, 20, 30, 40, and 50 m on Davies Reef during June.

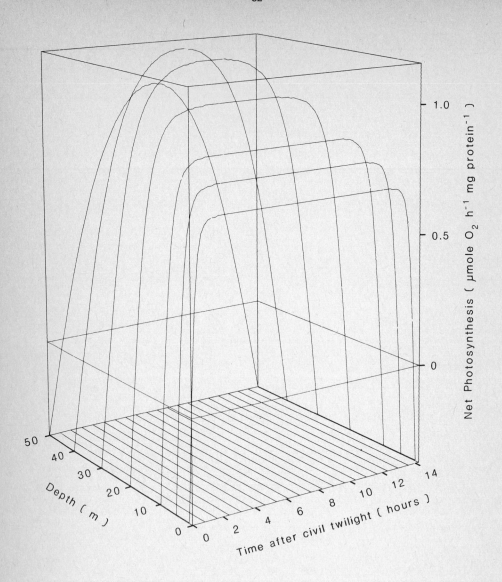

Figure 3. Estimated instantaneous *in situ* rates of net photosynthesis by *Acropora granulosa* at depths of 1, 5, 10, 20, 30, 40, and 50 m on Davies Reef during December.

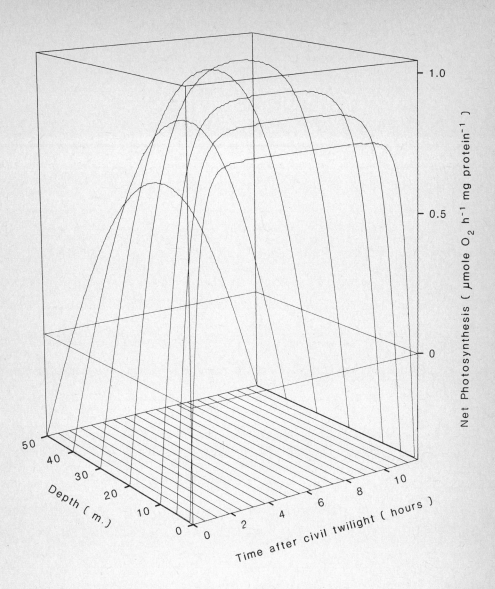

Figure 4. Estimated instantaneous <u>in situ</u> rates of net photosynthesis by <u>Acropora granulosa</u> at depths of 1, 5, 10, 20, 30, 40, and 50 m on Davies Reef during June.

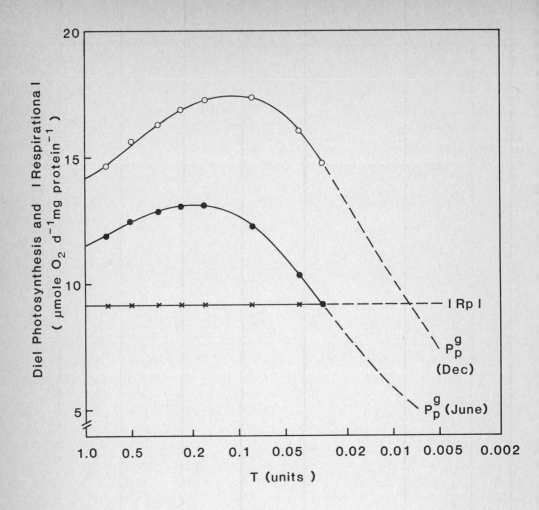

Figure 5. Calculated diel respiration (x) and photosynthesis during December (o) and June (●) for _Acropora granulosa_ on Davies Reef. Symbols represent depths at which experimental corals were collected.

ACKNOWLEDGEMENTS

We thank Messrs M. Devereux and M. Susic for skilled technical assistance, Drs A. Dartnall, F. Gillan and D. Kinsey for editorial criticism and Dr J. Veron for coral identification. Ms Marietta Tyssen drafted the figures, and the masters and crew of the R/V Lady Basten provided logistic support.

LITERATURE CITED

Chalker BE and Dunlap WC (1983a) Seasonal changes in the kinetics of primary production by reef building corals at Davies Reef, Great Barrier Reef, Australia. In: Abstracts of the Pacific Science Association, 15th Congress. Dunedin, New Zealand, p 36.

Chalker BE and Dunlap WC (1983b) Primary production and photoadaptation by corals on the Great Barrier Reef. In: Baker JT, Carter RM, Sammarco PW and Stark KP (ed.) Proceedings of the inaugural Great Barrier Reef conference. JCU Press, Townsville, Australia, p 293.

Chalker BE, Dunlap WC and Oliver JK (1983) Bathymetric adaptations of reef-building corals at Davies Reef, Great Barrier Reef, Australia II. Light-saturation curves for photosynthesis and respiration. J Exp mar Biol Ecol. In press.

Chalker BE and Taylor DL (1975) Light-enhanced calcification, and the rate of oxidative phosphorylation in calcification of the coral Acropora cervicornis. Proc R Soc Lond B 190: 323-331.

Davies PS (1977) Carbon budgets and vertical zonation of Atlantic reef corals. In: Taylor DL (ed.) Proceedings of the third international coral reef symposium vol 1. Rosenstiel School of Marine and Atmospheric Science, University of Miami, Miami, Florida, p 391.

Davies PS (1980) Respiration in some Atlantic reef corals in relation to vertical distribution and growth form. Biol Bull 158: 187-194.

Falkowski PG and Dubinsky Z (1981) Light-shade adaptation of Stylophora pistulata, a hermatypic coral from the Gulf of Eilat. Nature 289: 172-174.

Goreau TF (1959) The physiology of skeleton formation in corals. I. A method for measuring the rate of calcium carbonate deposition by corals under different conditions. Biol Bull 116: 59-75.

Goreau TF and Goreau NI (1959) The physiology of skeleton formation in corals. II. Calcium deposition by hermatypic corals under various conditions in the reef. Biol Bull 117: 239-250.

Hamming RW (1971) Introduction to applied numerical analysis. McGraw Hill, New York.

Her Majesty's Nautical Almanac Office (1980) The nautical almanac 1981. Her Majesty's Stationary Office, London.

Kawaguti S (1937) On the physiology of reef corals. I. On the oxygen exchanges of reef corals. Palao Trop Biol Stn Stud 1: 187-198.

MacCaull WA and Platt T (1977) Diel variations in the photosynthetic parameters of coastal marine phytoplankton. Limnol Oceanogr 22: 723-731.

Muscatine L (1980) Productivity of zooxanthellae In: Falkowski PG (ed.) Primary productivity in the sea. Plenum Press, New York p 381.

Muscatine L and Cernichiari E (1969) Assimilation of photosynthetic products of zooxanthellae by a reef coral. Biol Bull 137: 506-523.

Muscatine L, Pool R and Cernichiari E (1972) Some factors influencing selective release of soluble organic material by zooxanthellae from reef corals. Mar Biol 13: 298-308.

Smith D, Muscatine L and Lewis D (1969) Carbohydrate movement from autotrophs to heterotrophs in parasitic and mutualistic symbiosis. Biol Rev 44: 17-90.

Steemann Nielsen E and Hansen VK (1961) Influence of surface illumination on plankton photosynthesis in Danish waters ($66°N$) throughout the year. Physiol Plant 14: 595-613.

Steemann Nielsen E and Jørgensen EG (1968) The adaptation of plankton algae. I. General part. Physiol Plant 21: 401-413.

Talling JF (1957) Photosynthetic characteristics of some freshwater diatoms in relation to underwater radiation. New Phytol 56: 29-50.

Vandermeulen JH, Davis ND and Muscatine L (1972) The effect of inhibitors of photosynthesis on zooxanthellae in corals and other marine invertebrates. Mar Biol 16: 185-191.

Vandermeulen JH and Muscatine L (1974) Influence of symbiotic algae on calcification in reef corals : critique and progress report. In: Vernberg WB (ed.) Symbiosis in the sea. University of South Carolina Press, Columbia, South Carolina, p 1.

Vollenweider RA (1965) Calculation models of photosynthesis depth curves and some implications regarding day rate estimates in primary production measurements. In: Goldman CR (ed.) Primary productivity in aquatic environments. University of California Press, Berkley, p 425.

Wells JM (1977) A comparative study of the metabolism of tropical benthic communities. In: Taylor DL (ed.) Proceedings of the third international coral reef symposium vol. 1. Rosenstiel School of Marine and Atmospheric Science, University of Miami, Miami, Florida, p 545.

Wethey DS and Porter JW (1976a) Sun and shade differences in productivity of reef corals. Nature 262: 281-282.

Wethey DS and Porter JW (1976b) Habitat-related patterns of productivity of the foliaceous reef coral, *Pavona praetorta* Dana. In: Mackie GO (ed.) Coelenterate ecology and behavior. Plenum Press, New York, p 59.

Yonge CM (1973) Coral reef project - Papers in memory of Dr. Thomas F. Goreau. 1. The nature of reef-building (hermatypic) corals. Bull Mar Sci 23: 1-15.

Zvalinskii VI, Leletkin VA, Titlyanov EA and Shaposhnikova MG (1980) Photosynthesis and adaptation of corals to irradiance. 2. Oxygen exchange. Photosynthetica 14: 422-430.

GENERAL FEATURES OF PHYTOPLANKTON COMMUNITIES AND PRIMARY PRODUCTION IN THE GULF OF NAPLES AND ADJACENT WATERS

D.MARINO, M.MODIGH and A.ZINGONE

Stazione Zoologica, 80121 Naples, Italy

I. Introduction

The Gulf of Naples and that of Salerno are adjacent coastal embayments with differing basin morphology which open onto the oligotrophic Tyrrhenian Sea (fig. 1). The Gulf of Salerno is open with full access to external waters and has an average depth (260 m) greater than the Gulf of Naples (170 m). The latter Gulf has a more diversified coastline and depth profile. Excluding the Sorrentine peninsula, these regions are characterized by a high anthropogenic load from urban, industrial and agricultural origins. This is particularly true for the Gulf of Naples which is one of the most densely populated areas of the world.

The highly diversified morphology in addition to social-economic structures of these regions produces a strikingly complex hydrographic and biological situation. Moreover, the rapid transition existent from inner coastal to open ocean waters within the both Gulfs makes them a stimulating site for ecological research studies.

Most of the studies conducted at the Stazione Zoologica since its foundation (1872) regarded marine organisms of the Gulf of Naples from a morphological and physiological viewpoint. Therefore, only fragmentary ecological information was available until ten years ago (see Carrada and Rigillo Troncone, 1973, for a review) when research aimed at studying the Gulf of Naples as an integrated ecosystem was undertaken. More recently (1980-1983), such investigations have been extended to the adjacent waters of the Gulf of Salerno.

For the period 1975 to 1981, the sampling strategies and parameters measured in the Gulf of Naples were given by Carrada (1983). Subsequently, during the summer of 1983, standard hydrographic and biological parameters were measured weekly at stations located along the 50 and 100 m isobaths (fig. 1). The Gulf of Salerno was investigated for the first time during four major cruises in November 1981, April 1982, July 1982, December 1982 (fig. 1). Other phytoplankton samples were periodically collected at station C1 during summer (1980-1983).

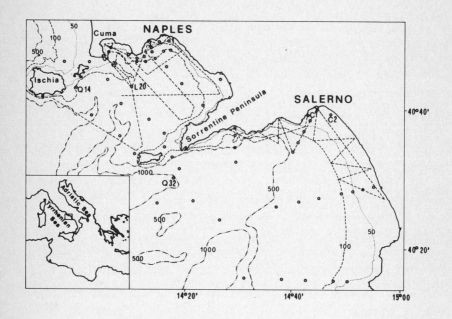

Fig. 1 - Map of the Gulfs of Naples and Salerno showing the location of surface sampling tracks (----) and stations (o).

The results of recent investigations are discussed here together with those previously published by Hopkins and GONEG (1977), Carrada et al. (1979a, 1979b, 1980, 1981a, 1981b, 1982), Marino and Modigh (1981) and Tomas (1981).

II. Results and discussion

Two different mechanisms for nutrient enrichment of the euphotic zone occurred in the Gulf of Naples and Salerno. Offshore waters having the same characteristics described for the Tyrrhenian Sea (Hopkins, 1978) depended on the vertical transport of deep water nutrients during winter isothermy. In contrast, inshore waters were mainly influenced by land run-off from small rivers, as well as domestic and industrial discharges.

A. Open waters

The physical-chemical properties of the local water masses are summarized in table 1 (Hopkins and GONEG, 1977; Carrada et al., 1980).

Table 1. Characteristics of the local water masses.

	Depth	T°C	S°/₀₀	NO_3-N (μg-at l^{-1})	SiO_4-Si (μg-at l^{-1})
TSW	0-75m (summer only)	13.5-26.5	37.6-38.2	0.05-0.3	1.0-3.0
TIW	75-100m (summer) 0-150m (winter)	13.6-14.2	37.8-38.6	0.1 -1.0	0.5-2.5
LIW	200-700m (summer) 300-900m (winter)	13.7-14.2	38.6-38.8	1.5 -4.0	0.5-6.0

TSW = Tyrrhenian Surface Water; TIW = Tyrrhenian Intermediate Water; LIW = Levantine Intermediate Water.

It is worth noting that the relatively high nutrient content of the LIW was strongly diluted by the overlaying oligotrophic TIW upon reaching the euphotic zone. In fact, the winter water column was homogeneous only down to about 150 m since water cooling was not adequate to cause mixing with underlaying layers. Nutrient levels and phytoplankton biomass in the euphotic zone remained low throughout the year.

Seasonal variations in species abundance and composition were similar in offshore areas of both Gulfs. In these regions phytoplankton populations showed a single abundance maximum (0.4-0.6 mg Chl a m^{-3}) at the beginning of March mainly due to diatoms, in particular *Asterionella glacialis, Thalassiothrix frauenfeldii, Rhizosolenia alata* f.*gracillima, Nitzschia closterium* and several species of *Chaetoceros* (*C. affine, C. curvisetum, C. decipiens, C. breve*). An increased vertical stability of the water column, a depletion of nutrients and a presumed intense grazing activity of zooplankton after the spring diatom maximum resulted in lower phytoplankton biomass levels (about 0.2 mg Chl a m^{-3}) as well as a change in species composition. During this period the bulk of the population consisted of small flagellates (<10 μ) and naked dinoflagellates. The annual minimum (0.01-0.15 mg Chl a m^{-3}) occurred in summer and characterized a less diversified phytoplankton community dominated by small flagellates. Coccolithophorids and dinoflagellates were also consistently present during this period, each of which comprised about 20% of the total population. *Emiliania huxleyi, Calcidiscus leptoporus, Prorocentrum balticum* and *Oxytoxum variabile* were frequently recorded. Relatively stable values of chlorophyll a (0.1-0.3 mg m^{-3}) were recorded during the autumn-winter period. Phytoplankton populations primarily consisted of coccolithophorids (*Emiliania huxleyi, Anthosphaera quadricornu, Syracosphaera pulchra, Umbellosphaera tenuis*, ecc.) and various diatoms. The latter group increased in abundance during winter when water column was well mixed.

Notwithstanding the low phytoplankton standing crops of these regions, the coccolithophorids were consistently abundant (up to 60%) throughout the year. In particular, the ubiquitous, opportunistic species *Emiliania huxleyi* was often dominant presumably due to its relatively high growth rate under a wide range of temperature-light conditions.

In winter the depth of the euphotic zone (about 2 μE m^{-2}sec^{-1}) was 70-80 m which extended to more than 100 m (about 10 μE m^{-2}sec^{-1}) in summer, suggesting that viable phytoplankton populations could occur in deep waters. This phenomenon, already recorded in the Western Mediterranean (Cahet et al., 1972; Estrada, 1981; Velasquez, 1981), was often observed in offshore waters of both Gulfs. Chlorophyll a and oxygen maxima were recorded below the thermocline in late summer and in autumn (fig. 2). Further studies are in progress to evaluate the contribution of deep water phytoplankton to the productivity of the whole water column.

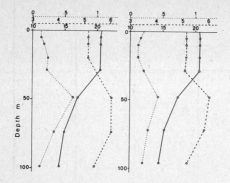

Fig. 2 - Vertical distribution of temperature (T °C ———), oxygen (O_2 ml l^{-1} ---) and chlorophyll a (mg m^{-3} ····) at two offshore stations (Q14, Q32) in October 1977.

B. Coastal waters

Inshore areas manifested unpredictable spatial and temporal variability in physical--chemical parameters contrasting offshore areas characterized by low phytoplankton biomass and similar species composition throughout the year for both Gulfs. Phytoplankton abundance, species composition and succession not only differed from those observed in open waters but also varied from one coastal area to another within each Gulf.

As mentioned previously, nutrient enrichment in coastal waters was mainly due to runoff and discharges scattered along the coast of both Gulfs. The effects of such nutrient inputs varied throughout the year in relation to climatic conditions as well as to the water column structure. In fact, highest biomass was recorded in surface layers during late spring and summer when stratification of water column limited the vertical diffusion of nutrient plumes. For the rest of the year, and particularly during well mixed winter conditions, nutrients were diluted in larger volumes of water and spread further offshore.

In the Gulf of Naples, intense and diversified sources of land runoff entering into a morphologically complex basin resulted in steep inshore-offshore gradients. Such gradients were observed simultaneously in different sites (Carrada et al., 1980). During summer, phytoplankton abundances in coastal waters often assumed values exceeding two orders of magnitude than those recorded for offshore waters. In particular, several phytoplankton blooms were observed along a narrow strip (about 2 Km) off Naples in July-August 1983. The spatial distribution and temporal succession of these blooms may reflect the small-scale heterogeneity of nutrient enrichment. Figure 3 shows the uncoupled fluctuations in chlorophyll a values recorded at two nearby stations. The core of the biomass was confined to surface waters where chlorophyll a values up to 50 mg m^{-3} and primary production values up to 2 g C m^{-3}d^{-1} were recorded. A rapid reduction in light intensity, with extinction coefficients reaching 0.4, was observed for the first five meters, restricting the euphotic zone to a depth of 10-15 m.

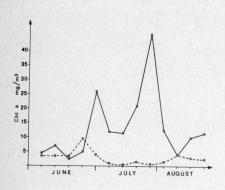

Fig. 3 - Variations in surface chlorophyll a values at stations 1 (——) and 3 (----) in the summer of 1983.

Cell numbers commonly varied from 10 to 40 million cells l^{-1}, but higher values up to 120 million were also recorded. Bloom populations included several species of diatoms (*Chaetoceros simplex, Leptocylindrus danicus, L. minimus, Nitzschia closterium, Skeletonema costatum, Thalassiosira decipiens*) together with *Emiliania hyxleyi* and *Eutreptiella* sp.. These species alternately dominated the phytoplankton whereas dinoflagellates and small flagellates rarely exceeded 20% of the total cell numbers. Values for species diversity from 1.70 to 3.04 ($\bar{x}$ 2.36) were mainly due to the concurrent dominance of more than one species.

Factors controlling the structure and dynamics of these populations are not well understood. The observed values for species diversity and the frequent dominance of small, fast-growing diatoms could be related to intense nutrient inputs in addition to wind driven turbulence of surface waters. Both factors could contribute in maintaining the system in a juvenile although diversified phase (Margalef, 1978).

Annual minima for chlorophyll a values for the Gulf of Naples (about 1 mg m^{-3}) were observed in late autumn or early winter, when disruption of the stratified water column occurred. The onset and duration of this period varied from year to year depending on weather conditions. After complete mixing, phytoplankton showed a more uniform spatial distribution reflecting the greater diffusion and dilution of nutrients derived from land runoff. In February 1979 highest surface chlorophyll a values (about 2 mg m^{-3}) were recorded in a vast area near the Procida and Ischia channels where waters influenced by the Volturno river plume and the Cuma sewage outfall occurred. The phytoplankton community (480 · 10^3 cells l^{-1}) was largely dominated by diatoms that represented 90% of the entire population (Carrada et al., 1981a).

The Gulf of Salerno as compared to the Gulf of Naples had less notable hydrographic and biological gradients giving a less marked spatial heterogeneity. Small rivers, that drain intensively cultivated fields south of Salerno, and urban sewage outfalls constituted the main sources of nutrient enrichment in these coastal waters. Maximal nutrient concentrations never attained those observed for coastal waters in the Gulf of Naples. This may be primarily due to differences in nutrient inputs and mixing dynamics for the two basins. Lower and more uniformly distributed values for phytoplankton biomass were recorded for inshore waters of the Gulf of Salerno.

Chlorophyll a concentrations ranging from 0.02 to 4.33 ($\bar{x}$ 0.65) mg m^{-3} were measured in surface coastal waters out to the 50 m isobath during July 1982. The highest values were recorded near the city of Salerno and in proximity to the outlet of a small river located in the southern part of the Gulf. Weighted mean values for primary production at two inshore stations (C1 and C2) were 11 and 6 mg C m^{-3}d^{-1}. Phytoplankton populations consisting mainly of dinoflagellates (small naked dinoflagellates and several species of *Protoperidinium* and *Prorocentrum*) and small flagellates less than 10 µ dominated the community. Diatom or dinoflagellate dominated blooms occasionally took place off the coast of Salerno in late spring and summer. Intense dinoflagellate blooms of

Gonyaulax cf. *grindleyi* (40 million cells l^{-1}) comprising 90% of the population were observed in July-August of 1980 and 1981. Such blooms were probably related to the prolonged stable weather conditions augmented by intense dredging activity in the Salerno Harbour. The dredge sediments deposited near the coast may have enriched surface waters with nutrients and could have released resting cysts.

In autumn, inshore phytoplankton concentrations only slightly exceeded those recorded in open waters (Carrada et al., 1982). Chlorophyll a values from 0.5 to 0.8 mg m^{-3}, corresponding to $30-100 \cdot 10^3$ cells l^{-1} and weighted mean primary production values of about 15 mg C $m^{-3} d^{-1}$ were measured. Inshore phytoplankton communities dominated by *Emiliania huxleyi* were similar to those observed in offshore areas, with the exception of a few stations situated near the outlets of small rivers. These stations were commonly dominated by small flagellates and diatoms.

C. The boundary regions

Major changes for physical-chemical and biological parameters defined the boundary region between inshore and offshore waters. The distance from the coast for this region varied throughout the year in relation to water circulation patterns and basin morphology.

Coastal waters in the Gulf of Naples extented approximately to the 100 m isobath in summer but extended to greater areas during the rest of the year (Carrada, unpubl.data). Such variations were best demonstrated by the annual phytoplankton cycle for station L20 located at the 300 m isobath (Carrada et al., 1979b). Biomass values (0.3-0.9 mg Chl a m^{-3}) and abundances of $30-200 \cdot 10^3$ cells l^{-1} were higher than those recorded for offshore areas, suggesting an input of diluted nutrients from inshore waters. Phytoplankton composition, however, was similar to that observed offshore for the greater part of the year. A peak in abundance was observed in autumn due to the overlapping of coastal and offshore communities. Primary production (Tomas, 1981) varied from 1.4 to 112 mg C $m^{-3} d^{-1}$.

Coastal waters in the Gulf of Salerno were generally restricted to nearshore areas and offshore waters in most cases bordered near the coast. After heavy rainfalls, influence of land runoff probably extended further offshore but unfortunately no data are available for these rainy periods.

III. Concluding remarks

The considerable spatial and temporal variability of phytoplankton communities in the Gulfs of Naples and Salerno makes it difficult to generalize these data for other coastal areas. This difficulty, a common feature for the Mediterranean as a whole, is due to the complexity of the basin, hydrological regimes and variability in anthropogenic inputs. Although available information does not allow for the formulation of a general model, a comparison can be made between the existing data and those recorded for other Mediterranean areas. With the exclusion of the Adriatic Sea (Gilmartin and Revelante, 1980), phytoplankton biomass and production values for the Eastern Mediterranean are generally lower than those recorded in our region. In fact, the Eastern Mediterranean is considered an oligotrophic basin with low phytoplankton biomass and production (Sournia, 1973; Becacos-Kontos, 1977) accompanied by low zooplankton biomass (Scotto di Carlo and Ianora, 1983).

Our data for phytoplankton biomass and production generally fell within or slightly above the values recorded for the Western Mediterranean (see for instance Castellvì and Ballester, 1981; Estrada, 1981; Kim, 1980; Magazzù, 1980). However those values measured in the vicinity of urban settlements were unusually high for the Mediterranean. In particular, data reported for eutrophied (Benon et al., 1977) and upwelling areas (Jacques et al., 1973) for the North Western Mediterranean never reached the maxima observed during the summer blooms in the Gulf of Naples nor in the Gulf of Salerno.

However, even when data of the Western Mediterranean fell within the same range, the driving forces determining the structure and production of phytoplankton communities were so complex that any comparison appears at this time haphazardous.

References

Becacos-Kontos T (1977) Primary production and environmental factors in an oligotrophic biome in the Aegean Sea. Mar.Biol. 42: 93-98.

Benon P, Blanc F, Bourgade B, David P, Kantin R, Leveau M, Romano JC and Sautriot D (1977) Impact de la pollution sur un écosystème méditerranéen côtier. II. Relations entre la composition spécifique des populations phytoplanctoniques et les taux de pigments et de nucléotides adényliques (ATP,ADP,AMP). Int.Revue ges.Hydrobiol. 62(5): 631-648.

Cahet G, Fiala M, Jacques G and Panouse M (1972) Production primaire au niveau de la thermocline en zone néritique de Méditerranée Nord-Occidentale. Mar.Biol. 14: 32-40.

Carrada GC (1983) The Gulf of Naples and its data base. In: Carrada GC, Hopkins TS, Jeftic L and Morcos S (eds) Quantitative analysis and simulation of Mediterranean coastal ecosystems: the Gulf of Naples, a case study. UNESCO Rep. Mar.Sci.20: 70-79.

Carrada GC and Rigillo Troncone M (1973) Indicazioni bibliografiche per una valutazione della evoluzione fisico-chimica e biologica del Golfo di Napoli. Fondazione Politecnica per il Mezzogiorno d'Italia. Quad. n. 70.

Carrada GC, Fresi E, Marino D, Modigh M and Ribera D'Alcalà M (1981a) Structural analysis of winter phytoplankton in the Gulf of Naples. J. Plankton Res. 3(2): 291-314.

Carrada GC, Hopkins TS, Bonaduce G, Ianora A, Marino D, Modigh M and Scotto di Carlo B (1980) Variability in the hydrographic and biological features of the Gulf of Naples. P.S.Z.N.I.: Marine Ecology 1: 105-120.

Carrada GC, Marino D, Modigh M and Ribera D'Alcala M (1979a) On the distribution of Utermöhl phytoplankton in a coastal sub-area of the Gulf of Naples. Rapp.Comm.int.Mer Médit. 25/26(8): 73-74.

Carrada GC, Marino D, Modigh M and Ribera D'Alcalà M (1979b) Observations on the annual cycle of Utermöhl phytoplankton at a fixed station in the Gulf of Naples. Rapp.Comm.int.Mer Médit. 25/26(8): 75-76.

Carrada GC, Marino D, Modigh M and Ribera D'Alcalã M (1981b) Variazioni spaziali in acque superficiali di nutrienti, clorofilla ed associazioni fitoplanctoniche nel Golfo di Napoli. Quad.Lab.Tecnol.Pesca 3(1 suppl.): 419-434.

Carrada GC, Marino D, Saggiomo V and Zingone A (1982) Popolamenti fitoplanctonici e condizioni ambientali nel Golfo di Salerno. Boll.Mus.Ist.Biol.Univ.Genova 50 suppl. 139.

Castellvì J and Ballester A (1981) Aspectos microbiologicos del estudio oceanografico de la plataforma continental II. Hidrologia y productividad primaria. Inv.Pesq. 45(2): 359-389.

Estrada M (1981) Biomasa fitoplanctonica y produccion primaria en el Mediterraneo occidental, a principios de otono. Inv.Pesq. 45(1): 211-230.

Gilmartin M and Revelante N (1980) Nutrient input and the summer nanoplankton in the Northern Adriatic Sea. P.S.Z.N. I: Marine Ecology 1: 169-180.

Hopkins TS (1978) Physical processes in the Mediterranean basins. In: Kjerfve B (ed.) Estuarine transport processes. Univ.of South Carolina Press; 269-310.

Hopkins TS and GONEG (1977) The existence of Levantine Intermediate Water in the Gulf of Naples. Rapp.Comm.int.Mer Médit. 24(2):39-41.

Jacques G, Minas HJ, Minas M and Nival P (1973) Influence des conditions hivernales sur les productions phyto- et zooplanctoniques en Méditerranée Nord-Occidentale. II.Biomasse et production phytoplanctonique. Mar.Biol. 23: 251-265.

Kim K-T (1980) Contribution à l'étude de l'écosystème pélagique dans les parages de Carry-le-Rouet (Méditerranée nord-occidentale). 3. Composition spécifique, biomasse et production du microplancton. Tethys 9(4): 317-344.

Magazzù GA (1980) Primary production cycle in the South Italian coastal seas. Tethys 9(3): 207-213.

Margalef R (1978) Life-forms of phytoplankton as survival alternatives in an unstable environment. Oceanol. Acta 1(4): 493-509.

Marino D and Modigh M (1981) An annotated check-list of planktonic diatoms from the Gulf of Naples. P.S.Z.N. I: Marine Ecology 2(4):317-333.

Scotto di Carlo B and Ianora A (1983) Standing stocks and species composition of Mediterranean zooplankton. In: Carrada GC, Hopkins TS, Jeftic L and Morcos S (eds) Quantitative analysis and simulation of Mediterranean coastal ecosystems: the Gulf of Naples, a case study. UNESCO Rep.Mar.Sci. 20: 59-69.

Sournia A (1973) La production primaire planctonique en Méditerranée. Essai de mise à jour. Newsletter of the Cooperative Investigations in the Mediterranean 5(1): 1-128.

Tomas CR (1981) Primary production in the Gulf of Naples: winter-spring 1980. Rapp.Comm.int.Mer Médit. 27(7): 67-68.

Velasquez ZR (1981) Summer phytoplankton in the Catalan Sea. Rapp.Comm. int. Mer Médit. 27(7): 79-81.

UNDERSTANDING OLIGOTROPHIC OCEANS : CAN THE EASTERN MEDITERRANEAN BE A USEFUL MODEL ?

T.BERMAN[1], Y.AZOV[1] and D.TOWNSEND[2]

[1] Israel Oceanographic & Limnological Research Co.,
P.O.Box 8030, Haifa, Israel

[2] Bigelow Laboratory for Ocean Sciences, West Boothbay Harbor,
Maine 04575, USA.

This paper will pose some general questions concerning marine ecosystems in the light of major discoveries and observations in the past decade. More specifically we shall apply these to oligotrophic seas and present some preliminary data to support the suggestion that the Eastern Mediterranean Basin may serve as a convenient model for the study of such environments.

The following points should be considered (not necessarily in order of precedence).

1) It has become increasingly evident that a very significant portion of phytoplankton biomass and photosynthetic activity in the open ocean, especially in oligotrophic regions, is associated with organisms smaller than 3 µm or even 1 µm (Johnson and Sieburth 1979; Waterbury et al. 1979; Li et al. 1983). Many of these small phytoplankters are blue-greens (Cyanobacteria), an algal division which was previously believed to be rather poorly represented in the pelagic flora. Often these ultra (or pico) phytoplankton are relatively more numerous towards the bottom of the euphotic zone or in deep chlorophyll maxima (see below) and there is some evidence that they are adapted to low light in the green region of the spectrum (Platt et al. 1983; Glover et al. 1984). Fundamental, unresolved, questions concerning these organisms are: what are their growth rates in situ and what organisms are capable of grazing on them?

2) The possible functions of the heterotrophic protozoa, until recently a comparatively neglected component of the plankton, have been attracting increasing research scrutiny (Banse 1982; Sherr and Sherr 1983). Microciliates such as tintinnids have been found to be major grazers of phytoplankton in some areas (Rassoulzadegan and Etiénne 1981; Capriulo and Carpenter 1983) and the smaller microflagellates may be important consumers of bacteria and ultraphytoplankton (Fenchel 1982; Sherr and Sherr 1983; Sherr et al. 1983). As yet, however, there are few reliable data on the amounts of carbon biomass which are grazed by the heterotrophic microzooplankton or the extent to which these organisms in their turn serve as prey. Growth efficiencies are a basic but as yet, an almost unknown parameter which are required for understanding the ecological roles of these protozoans.

In addition to their function as grazers, it seems probable that these organisms are important in the regeneration of nutrients such as ammonia and phosphorus (Sherr et al. 1983). They may also selectively facilitate the breakdown of specific detrital components such as polysaccharides (Sherr et al. 1982) and maintain elevated heterotrophic bacterial metabolism by actively cropping bacteria (Sieburth and Davis 1982).

3) Our conceptions regarding the functions of bacteria in the marine ecosystem have also undergone revision (Pomeroy 1980; Williams 1983). In the upper waters of the ocean most of the heterotrophic activity appears to be associated with single rather than with clumped or attached bacteria. The amount of carbon flux from algal extracellular release of organic matter to the heterotrophic bacteria remains controversial (Sharp 1977), but at least in some cases this may be a considerable portion of the photosynthetic fixation (Smith 1982). Despite some innovative methodological approaches, in situ growth rates of bacteria have remained difficult to evaluate. Since bacterial cell numbers seem to be rather constant in the upper ocean layers (Williams 1983) what then is the turnover time of carbon at this trophic level? Is being eaten by a microprotozoan a likely fate for a marine bacterium?

4) Deep chlorophyll layers (DCL), usually located at about 1% of the incident surface light are a common feature in many seas and their formation may be due to a variety of causes (Cullen 1982). A DCL may not necessarily represent a region of greater phytoplankton biomass but may only reflect the increase of chlorophyll content per cell due to the low level of irradiance. Venrick (1982) has provided detailed evidence to show that, in the North Pacific Central Gyre, qualitatively different algal populations were found in the DCL and in the upper phototrophic zone. The extent to which the phytoplankton are actively growing in DCL's and their contribution to overall ecosystem and energy transfer remain unresolved.

5) The existence of microscale inhomogeneities of chemical concentrations in a turbulent environment may have profound biological and ecological importance. One aspect of this is the suggestion that, in nutrient limited environments, the phytoplankton might derive some or all of their required nutrition in pulses from zooplankton excretions (McCarthy and Goldman 1979; Lehman and Scavia 1982). This conjures up the rather poignant picture of hungry algae in oligotrophic waters drifting around waiting for a handout! It also implies that the ability to assimilate a nutrient rapidly confers a distinct competitive advantage. Thus there may be two kinds of algal strategists, those with uptake systems which maximize for substrate affinity and those adapted for maximum uptake rate.

Bell et al (1974) proposed the idea of a 'phycosphere' zone of dissolved organic compounds around algal cells in which heterotrophic bacteria could presumably develop on the readily available substrates which are being released. Recently Azam (1983) has postulated that bacteria may in fact be actively attracted to such 'phycospheres' by biochemical 'signals' from compounds such as cyclic-AMP. Experimental evidence is still necessary to confirm these suggestions.

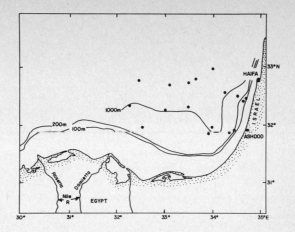

Fig. 1. General area of AID cruises. Only pelagic stations are marked.

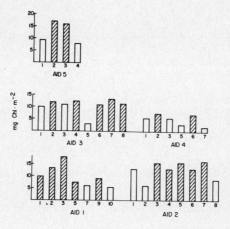

Fig. 2. Areal concentrations of chlorophyll (mg Chl.m^{-2}).
☐ neritic ▨ pelagic stations.

Table 1. Secchi depths, diffuse downwelling attenuation coefficients and chlorophyll.

Station	Secchi depth (m)	K (m^{-1})	Chl*
AID 1-1	42.0	-	0.026
1-2	9.5^x	-	0.210
1-3	12.5^x	0.112	0.093
1-5	37.0	0.047	0.032
1-7	6.0^x	0.230	0.222
1-9	41.0	0.040	0.024
1-10	15.0^x	0.080	0.133
AID 2-1	27.0	0.075	0.104
2-2	18.0^x	0.070	0.088
2-4	45.0	0.046	0.069
2-6	45.0	0.044	0.053
2-8	21.0	0.073	0.118
AID 3-1	18.0^x	0.057	0.245
3-5	23.0	0.058	0.079
3-6	33.0$^+$	-	0.031
3-7	39.0	0.034	0.039
AID 4-1	27.0	0.069	0.052
4-2	46.0	0.039	0.023
4-3	20.0^x	0.103	0.104
4-4	41.0	0.039	0.029
4-5	18.5^x	0.092	0.107
4-6	36.0	-	0.039
4-7	20.0^x	0.074	0.032
AID 5-1	36.0	-	0.094
5-2	30.0	0.054	0.136
5-4	27.0	0.053	0.120

*Chlorophyll (mg.m^{-3}) average of integrated values to 37% surface light.
$^+$Stormy conditions - probably underestimated. ^{x}Neritic stations.

Table 2. Relations between near-surface chlorophyll (Chl), Secchi depths (S) and diffuse downwelling attenuation coefficient (K).

1. Regression of Chl on K
   Chl = 0.837 K + 0.0314   r = 0.59   n = 21 (P 1%)

2. Regression of Chl on S
   Chl = -0.0039 S + 0.1975   r = -0.73 n = 26 (P 0.1%)

3. Regression of K on S
   K = -0.0027 S + 0.1435   r = 0.74   n = 21 (P 1%)

4. Regression of 1/S on Chl (Megard et al. 1980)
   1/S = 0.371 Chl + 0.013   r = 0.73   n = 26 (P 0.1%)

Obviously the above topics do not exhaust the roster of important unresolved questions concerning marine ecosystems but they illustrate the point that any comprehensive paradigm will need to encompass many new observations. In respect to oligotrophic seas, a brisk controversy still rages which can be summarized thus: 1) Are the primary producers in the oligotrophic seas growing at 1-2 doublings day^{-1} or at 0.2 doublings day^{-1}? 2) Is there greater biomass in the higher trophic levels than can be accounted for by primary productivity as measured by conventional means in these regions?

THE EASTERN MEDITERRANEAN AS A MODEL OLIGOTROPHIC SEA.

In 1981 we began a baseline study on the biological productivity of the Eastern Mediterranean Basin. Our initial observations prompt us to suggest that this may be a very convenient area in which to study the structure of an extremely oligotrophic ecosystem. Here we present some data on optical characteristics, chlorophyll distribution and primary productivities to substantiate this claim.

Hydrostations and Methods

The data were collected during five cruises (AID-1, 23-31 July 1981; AID-2, 10-14 December 1981; AID-3, 9-12 April 1982; AID-4, 18-22 July 1982; AID-5, 13-15 February 1983), covering the area shown in Fig. 1. Surface chlorophyll was monitored between stations en route by in vivo fluorescence (Lorenzen 1966). All water samples from hydrocasts were prefiltered through a 130 μm mesh Nitex net. For size distribution studies, samples were subsequently filtered through 10 μm Nitex nets or 3 μm Nuclepore filters. Chlorophyll and phaeophytin concentrations were measured in discrete samples fluorometrically after concentration on 0.45 μm Millipore filters and extraction in 90% acetone (Holm-Hansen et al. 1965). Photosynthetic carbon fixation was measured with ^{14}C bicarbonate using an on-deck simulator cooled with surface water with incubations of 4 to 6.5 hours. Photosynthetically available surface and downwelling irradiance were determined with a LiCor Quantum meter.

Transparency and downwelling attenuation coefficients

In Table 1 we list Secchi depths, diffuse downwelling attenuation coefficients (K) and chlorophyll concentrations integrated to approximately 37% of surface irradiance. (This corresponds to one attenuation unit and is approximately the depth to which chlorophyll concentrations are determined by remote sensors such as the Coastal Color Zone Scanner mounted on the Nimbus 7 satellite).
Secchi depths ranged from maxima in the offshore waters of 42 to 46 m (in July 1981 and 1982 respectively) to a minimum of 6 m for very close inshore waters (July

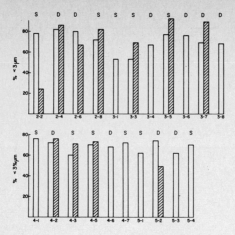

Fig. 3. Percentage of areal chlorophyll (☐) or primary productivity (▨) associated with organisms <3 μm. Nearshore (S) and pelagic (P) stations.

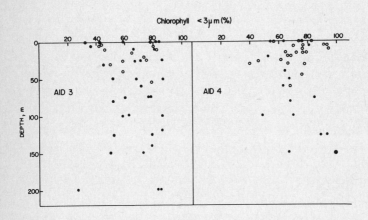

Fig. 4. AID 3 and AID 4: Depth distribution of chlorophyll associated with organisms <3 μm. Nearshore (○) and pelagic (●) stations.

1981). Profiles of diffuse downwelling light attenuation , K, exhibited classical exponential decreases with depth although we always recorded relatively higher values of K in the water layers from 0 to 5 or 10 m. This phenomenon was probably not due to greater pigment attenuation but was mainly the result of the large changes in the spectrum of light energy which occur near the surface (see Jewson et al. 1984).

Although plant pigments are only partly responsible for the light attenuation in the sea, both Secchi depths and K were significantly correlated with the chlorophyll concentrations down to about 37% (Table 2). The regression of chlorophyll on Secchi depths was the most significant (0.1% level). We suggest that further investigation may show that Secchi measurements can be extremely useful in this region for providing rapid estimations of nearsurface chlorophyll and light attenuation which are required to validate data acquired by remote sensing.

The Secchi depths and the values of the diffuse downwelling coefficients in the pelagic waters were typical for oligotrophic seas. It should be noted however that neither the Secchi depth values nor those of the attenuation coefficients reflect the contribution of deep chlorophyll maxima to the total areal standing crop of this pigment.

Chlorophyll

The areal concentrations of chlorophyll in the euphotic zone ranged from 3.1 to 18.0 mg.m^{-2} (Fig. 2). Note that while the average chlorophyll concentration (mg.m^{-3}) at pelagic stations ($>$80 m) was much lower than that at neritic sites ($<$80 m) the total chlorophyll standing stocks (mg.m^{-2}) in the former locations were frequently greater. The lowest areal chlorophyll concentrations for neritic waters were observed in AID-4 (July 1981).

At all seasons, both for nearshore and pelagic waters; the majority of the chlorophyll was associated with organisms smaller than 3 μm (Fig. 3). However there was no apparent trend to greater proportions of picoplankton with depth (Fig. 4), as has been reported elsewhere (Li et al. 1983).

For all pelagic stations of cruises AID-1 through 4, we observed deep chlorophyll maxima at depths from 75 to 150 m (Fig. 5). The two 'deep' stations measured in AID-5 (5-2, 5-3) were exceptions to this. In future work it will be of importance to investigate whether these deep chlorophyll maxima are in any way seasonal phenomena.

As yet we have not completed the interpretation of the in vivo fluorometric surface scans from all cruises. During AID-1 a sharp front was observed both in chlorophyll and surface temperatures between the nearshore and pelagic waters. On other cruises (e.g. AID-5) the surface front was absent although in general a gradient from higher to lower surface chlorophyll was recorded over the first 20 kms as the vessel steamed west from the shore.

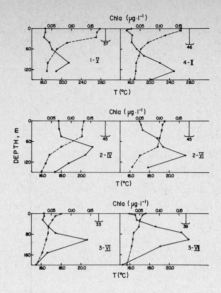

Fig. 5. Depth distribution of chlorophyll (●——●), temperature (●---●) and Secchi depth (m) at various pelagic hydrostations.

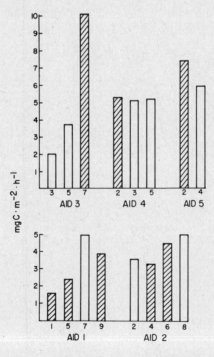

Fig. 6. Areal primary productivities (mg $C.m^{-2}.h^{-1}$) ☐ neritic ▨ pelagic stations.

Primary Productivity

The measurement of phytoplankton primary production using on-deck simulation with ^{14}C methodology has been subjected to considerable criticism in recent years (Gieskes et al. 1979· Eppley 1980). We emphasize, therefore, that our results must be viewed with reservation until they can be substantiated by other techniques.

In our study the measured areal photosynthetic rates ranged from 1.62 to 10.2 $mgC.m^{-2}.h^{-1}$ (Fig. 6). Higher volumetric values ($mgC.m^{-3}.h^{-1}$) were observed in the richer nearshore waters but here total areal productivity was limited by depths which were frequently less than 1% light attenuation depth. Assimilation numbers averaged 2.2 (s.d. $\pm$ 2.1) $mgC.mgChl^{-1}.h^{-1}$.

Although accurate conversions from hourly to daily rates are not possible, rough estimations based on effective sunlight hours give average daily values of primary productivity of about 40 to 50 $mgC.m^{-2}.d^{-1}$. These would imply that the annual net photosynthetic carbon fixation in these waters might be somewhere between 10-20 $gC.m^{-2}.y^{-1}$. Differential filtration indicated that, as with chlorophyll, the majority of the photosynthetic activity was associated with organisms smaller than 3 µm (Fig. 3). However, the assimilation numbers of the picoplankton (<3 µm) were not generally higher than those of the larger phytoplankton, in contrast to other observations (Li et al. 1983).

Conclusions

Despite their limited scope, our results show a consistent picture. The pelagic waters of the Mediterranean Basin are highly oligotrophic, having extremely low phytoplankton standing crops and apparent low rates of primary production as measured by standard ^{14}C methodology. Their transparency and optical characteristics reflect the low algal populations. A deep chlorophyll maximum, usually at about 1% of incident light level, is generally observed. The nanoplankton (<20 µm); and more especially, the picoplankton (<3 µm), constitute the majority of the chlorophyll biomass and are responsible for most of the photosynthetic activity. The nearshore water mass roughly overlying the narrow (10-20 kms) Israeli coastal shelf, has much higher levels of chlorophyll standing crops and primary productivity and is often clearly delineated from the pelagic water by a surface front of temperature and chlorophyll.

In the first part of this paper we briefly reviewed some recent observations and discoveries which, when more fully investigated, will undoubtedly greatly influence our understanding of oceanic ecosystems. Our data indicate that the pelagic waters of the Eastern Mediterranean Basin has many features characteristically associated with nutrient poor oceans. This area, because of its relatively small size and the proximity of shore based research facilities, may prove particularly convenient for further studies which may elucidate some of the many mysteries still clouding our perception of the workings of oligotrophic seas.

Acknowledgements

This paper is dedicated to the memory of the late O.H. Oren, a pioneer of Israeli oceanographic research, who died on 4 November 1983. We thank Drs. A. Hecht and B. Kimor in Israel and Drs. N.M. Dowidar and S. Sharif El Din in Egypt for constructive comments and discussions. For sampling and technical assistance we are grateful to A. Shneller, Y. Almog, Hussein Abd El Reheim and to Captain A. Zur and crew of the R/V Shikmona. This research was funded by a grant from the U.S. Agency for International Development and is a contribution from the Israel Oceanographic & Limnological Research Co., and the Bigelow Laboratory for Ocean Sciences. No. 83028.

References

Azam F (1983) Bacterioplankton secondary production and its regulation by environmental factors. Proc. Workshop on Measurement of Microbial Activities in the Carbon Cycle in Aquatic Ecosystems. Plön (in press).

Banse K (1982) Cell volumes, maximal growth rates of unicellular algae and ciliates, and the role of ciliates in the marine pelagial. Limnol. Oceanogr. 27(6) :1059-1071.

Bell WH Land JM and Mitchell R (1974) Selective stimulation of marine bacteria by algal extracellular products. Limnol. Oceanogr. 19 : 833-839.

Capriulo GM and Carpenter EJ (1980) Grazing by 35 to 202 µm microzooplankton in Long Island Sound. Mar. Biol. 56 : 319-326.

Capriulo GM and Carpenter EJ (1983) Abundance, species composition, and feeding impact of tintinnid micro-zooplankton in Central Long Island Sound. Mar. Ecol. Prog. Ser. 10 : 277-288.

Cullen JJ (1982) The deep chlorophyll maximum: comparing vertical profiles of chlorophyll a. Can. J. Fish Aqua. Sci. 39 : 791-803.

Eppley RW (1980) Estimating phytoplankton growth rates in the central oligotrophic oceans. In: Falkowski PG (ed.) Primary Productivity in the Sea. Plenum Press, New York, p. 231-242.

Fenchel T (1982) Ecology of heterotrophic microflagellates. II. Bioenergetics and growth. Mar. Ecol. Prog. Ser. 8 : 225-231.

Gieskes WWC Kraay GL and Baars MA (1979) Current ^{14}C methods for measuring primary production : gross underestimates in oceanic waters. Netherland J. Sea Res. 13: 50-78.

Glover HE Phinney DA and Yentsch CS (1984) Photosynthetic characteristics of picoplankton compared with those of large phytoplankton populations in various water masses in the Gulf of Maine. Biolog. Oceanogr. (in press).

Holm-Hansen O Lorenzen CJ Holmes RW and Strickland JDH (1965) Fluorometric determinations of chlorophyll. J. Cons. Cons. Int. Explor. 30 : 3-15.

Jewson DH Talling JF Dring M Tilzer M Heaney I and Cunningham C (1984) Some problems caused by differences in spectral response and collecting properties of instruments. J. Plankton Res. (in press).

Johnson PW and Sieburth McN (1979) Chroococcoide cyanbacteria in the sea: a ubiquitous and diverse phototrophic biomass. Limnol. Oceanogr. 24 : 928-935.

Lehman JT and Scavia D (1982) Microscale patchiness of nutrients in plankton communities. Science 216 : 729-730.

Li WKW Subba Rao DV Harrison WG Smith JC Cullen JJ Irwin B and Platt T (1983) Autotrophic picoplankton in the tropical ocean. Science 219 : 292-295.

Lorenzen CJ (1966) A method for the continuous measurement of *in vivo* chlorophyll concentration. Deep Sea Res. 13 : 223-227.

McCarthy JJ and Goldman JC (1979) Nitrogenous nutrition of marine phytoplankton in nutrient-depleted waters. Science 203 : 670-672.

Megard RO Settles JR Bayer H and Combs WS (1980) Light, Secchi discs and trophic states. Limnol. Oceanogr. 25 : 373-378.

Platt T Subba Rao DV and Erwin B (1983) Photosynthesis of picoplankton in the oligotrophic ocean. Nature 301 : 702-704.

Pomeroy LR (1980) Microbial roles in aquatic food webs. In: Colwell RR (ed.). Aquatic Microbial Ecology, University of Maryland. p. 85-109.

Rassoulzadegan F and Etiénne M (1981) Grazing rate of the tintinnid *Stenosemella ventricosa* (Clap. and Lachm.) Jorg, on the spectrum of the naturally occurring particulate matter from a Mediterranean neritic area. Limnol. Oceanogr. 26 : 258-270.

Sharp JH (1977) Excretion of organic matter by marine phytoplankton. Do healthy cells do it? Limnol. Oceanogr. 22 : 381-399.

Sherr BF Sherr EB and Berman T (1982) Decomposition of organic detritus: A selective role for microflagellate protozoa. Limnol. Oceanogr. 27 : 765-769.

Sherr BF and Sherr EB (1983) Enumeration of heterotrophic microprotozoa by epifluorescence microscopy. Est. Coast. Shelf Sci. 16 : 1-7.

Sherr BF Sherr EB and Berman T (1983) Grazing, growth and ammonia excretion rates of a heterotrophic microflagellate fed four species of bacteria. Appl. Environ. Microbiol. 45 : 1196-1201.

Sieburth J McN and Davis PG (1982) The role of heterotrophic nanoplankton in the grazing and nurturing of planktonic bacteria in the Sargasso and Caribbean Sea. Annales inst. Oceanogr. 58(S) ; 285-296.

Smith REH (1982) The estimation of phytoplankton production and excretion by carbon-14 Mar. Biol. Let. 3 : 325-334.

Venrick EL (1982) Phytoplankton in an oligotrophic ocean: observations and questions. Ecol. Monographs. 52 : 129-154.

Waterbury JB Watson SW Guillard RRL and Brand LE (1979) Widespread occurrance of a unicellular marine planktonic cyanobacterium. Nature 277 : 293-294.

Williams PJLeB (1983) Bacterial production in the marine food chain: The Emperor's new suit of clothes? In: Flows of energy and materials in marine ecosystems:Theory and Practice. NATO-ARI, Bombannes (in press).

GROWTH RATES OF NATURAL POPULATIONS OF MARINE DIATOMS AS DETERMINED IN CAGE CULTURES

G.A. VARGO

Department of Marine Science, University of South Florida, 140, Seventh Ave S, St. Petersburg, Florida 33701, USA.

Introduction

The term 'cage' culture, as defined by Sakshaug and Jensen (1978), refers to " a population of living organisms retained behind a membrane which is permeable to the dissolved constituents of the growth medium as well as to the soluble metabolites excreted by the organisms."

Two types of cage culture chambers are commonly used; tubes made of dialysis membrane (Sakshaug and Jensen, 1978) and plexiglass chambers using Nuclepore$^{(R)}$ filters as cage walls (Owens et al., 1977). For simplicity, dialysis chambers will be called tubes and 'cages' will refer to filter-walled chambers.

Considering the interest in and the problems involved with estimating the growth and production rates of phytoplankton assemblages in oligotrophic, oceanic waters (Goldman et al.,1979; Goldman, 1980; Eppley, 1980,1981), cage cultures should be ideally suited for obtaining estimates of the potential _in situ_ growth rates of individual species in natural assemblages. Populations enclosed in a porous membrane and incubated _in situ_ or in tanks of flowing sea water allows them to react to "natural" spatial and temporal variations in temperature, salinity, nutrient concentrations and irradiance. The response of the entire assemblage and individual species can be obtained. No other method is available which can yield similar results. More commonly used methods e.g. ^{14}C, O_2, proximate analyses, measure the response of the community as a whole. While microautoradiography offers some potential for estimating the production rates of individual species the technique is even more laborious than using cage cultures. Estimates of growth rates for populations in oligotrophic, oceanic waters based on ^{14}C uptake requires answers to questions being raised about heavy metal and trace metal toxicity (Carpenter and Lively, 1980; Fitzwater et al., 1982), bottle size (Gieskes et al., 1979), and ultraviolet radiation (Smith and Baker, 1980). There are also problems associated with phytoplankton biomass measurements (Sakshaug, 1980) which are required to calculate growth rates from carbon or nitrogen uptake rates. Growth rate estimates based on uptake and biomass measurements also assume that balanced growth occurs, whereas Eppley (1981) argues that there is no reason to assume that balanced growth occurs in natural populations over short time scales.

Cage culture techniques offer an alternative method for establishing potential growth rates for natural phytoplankton populations. Growth rates greater than 2 divisions day^{-1} for individual species in dialysis culture of natural populations have been reported (Vargo, 1976,1979). Recently, Furnas (1982a) has reported that growth in filter-walled cages (modelled after those developed by Owens et al., 1977) was, at times, 2 to 3 times higher than for the same populations grown in dialysis tubing. Our results suggest that phytoplankton have the intrinsic ability to grow at rates which are higher than those normally reported in the literature (Goldman et al., 1979; Eppley, 1981) and that the type of cage may influence the maximum potential growth rate achieved by enclosed populations.

In this paper I will present some examples of growth of natural populations in tubes and cages, assess the reasons for expecting differences or similarities, report growth rates of several species from the eastern Gulf of Mexico and discuss these results with respect to current questions being raised about the growth rates of phytoplankton from oligotrophic, oceanic waters.

Methods

Dialysis tubes were 2.86 cm. in diameter and had a final, filled length of 24.8 cm. This size tube has a surface area of 229cm^2 with a volume of approximately 160 ml and an A/V ratio of 1.43. Replicate tubes were incubated in flowing sea water suspended in plexiglass tanks on the wheel arrangement described by Jensen et al.(1972). A small bubble in the tube provided continuous mixing. Cages were constructed according to the design of Owens et al. (1977) but were larger and used a 1 micron pore size, 142mm diameter Nuclepore$^{(R)}$ filter as cage walls. Two filters per cage were separated by a 2.54 cm thick plexiglass plate. Total surface area was 226 cm^2 but the total open pore area was 35.5 cm^2. Each cage contained 287 ml yielding a ratio of open pore area to volume of 0.12. Replicate cages were incubated in the same tank as the dialysis tubes, either mounted on the wheel or suspended within the tank. Furnas (1982a) found no difference in the growth rates of populations incubated in stationary and rotating cages.

Nitrogen transport rates were determined using dialysis tubes which were 1.6 cm in diameter, 25 cm long with a volume of 50 ml and an A/V ratio of 2.5. Ten replicate tubes were filled with filtered sea water to which ammonium or phosphate was added to achieve a final concentration of 5 μM. Two tubes were removed and the tank water was sampled at 15 min. intervals. Ammonium and phosphate concentrations were determined on a Technicon AutoAnalyzer$^{(R)}$. Transport rates were calculated from the following equation as modified by Schultz and Gerhardt (1969):

$$P_m = N/A_m \Delta S \qquad (1)$$

where P_m is the permeabilty coefficient (cm hr^{-1}), N is the rate of diffusion (g hr^{-1}); A_m is the total membrane area (cm^2) and ΔS is the concentration gradient (g cm^{-3}).

Growth rates based on cell counts are expressed as "k" (doublings day^{-1}, log$_2$). Rates based on chlorophyll$_a$ determinations (Holm-Hansen et al., 1965) are expressed as "μ" (doublings day^{-1}, log$_2$). Replicate cell counts of nearshore and Tampa Bay samples were made in a Sedgwick-Rafter counting chamber. Offshore samples were counted on a Zeiss, phase contrast, inverted microscope. The entire contents of the dialysis tube or cage was concentrated, rediluted to 10 ml and settled. The entire area of the settling chamber was counted.

Results and Discussion

Nutrient Transport Rates

Furnas (1982a) felt that both the lower yield and growth rates of selected species from natural populations in dialysis tubes could be explained by lower nutrient fluxes through the membranes compared to filter-walled cages. He derived phosphate exchange rates from laboratory experiments which were 3 to 6 times lower for dialysis membranes than for filter walled cages (0.093 ml cm^{-2} hr^{-1} and 0.3 to 0.6 ml cm^{-2} hr^{-1}, respectively). Nitrogen flux rates were not measured but were assumed to be equal to phosphate. Furnas (1982a) estimated that higher flux rates in cages would support approximately 3 times more chlorophyll than in dialysis tubes. This should also yield a longer exponential phase and possibly higher growth rates.

I have measured ammonia flux rates through dialysis membranes using the same size tubing as Furnas (1982a). He measured flux rates in beakers using deionized water and a 1 millimolar initial phosphate concentration gradient. My rates are, however, based on measurements made under the same experimental conditions used to incubate samples.

Decreases in ammonium and phosphate followed an exponential decay. Flux rates were calculated from the average difference between the tubes and the tank over a 1 hour period using the mean concentration gradient. My results indicate that the transport rate for ammonium was equivalent to that found by Furnas (1982a) for phosphate (Table 1). My estimate for phosphate flux is lower. Half times for ammonium flux are all less than 1 hour. The difference between the flux rates of nitrogen and phosphorus in dialysis tubes may be related to the size of their respective ions. The ionic radius of N^{3-} is approximately ½ that of P^{3-}, (1.71Å vs. 2.12Å respectively). Bond lengths and the charge of each ion will also influence its interaction with the membrane and the water molecules in the membrane pores, thereby affecting the actual diffusion rate. The permeation rate of a solute through a membrane follows Fick's Law. As such it is directly proportional to the concentration gradient and to the ratio of open pore area to pore length. Dialysis tubing, with a pore diameter of 48Å and a tortuous pathway 25 μm thick may have more interaction with diffusing solutes than a polycarbonate filter with a pore size of 10000 Å and a straight pathway of 10 μm. Therefore, I

TABLE 1: Transport rates and the half-time for diffusion of ammonium and phosphate through dialysis membranes.

Ion	ml $cm^{-2} hr^{-1}$	area, cm^2	ml hr^{-1}	t ½ (hr)
PO_4	0.04	170	6.3	5.5
NH_4^+	0.45	170	76.0	0.5
NH_4^+	0.59	170	99.0	0.4
NH_4^+	0.56	229	128.0	0.9

would expect higher flux rates through filters although the actual transport rates will, of course, be a function of the concentration gradient, mixing, and the interaction of solute ions with the membranes. My ammonium flux rates were based on a 5 µM difference in concentration gradient whereas Furnas's phosphate fluxes were based on an initial gradient of 1000 µM. Since our flux rates are essentially equivalent and since there were large differences in the concentration gradient and the pore area to volume ratio of the two types of cages, our transport rates may represent maximum values.

If they are maximum values then we can assume that the ammonium fluxes measured in dialysis tubes are comparable to filters. Dialysis tubes should, therefore, support a biomass equivalent to that in a cage before membrane transport becomes limiting and the population enters the linear growth phase. Growth rates should be equivalent.

Growth Rates

Comparisons between growth rates in tubes and cages were made using natural populations from Tampa Bay, Florida during the summer of 1982. Populations were screened through a 153 µm Nitex$^{(R)}$ net to remove larger zooplankton. Growth rates, estimated from changes in chlorophyll and cell numbers over a two day incubation period indicated there was little difference in growth rates using the two methods with 3 exceptions (Table 2). In 2 of these 3 exceptions growth rates in the cages were lower than in tubes and in one pair of cages where light was reduced by an additional 50%, the growth rate was higher.

Growth rates of some species from one trial (trial 2) were faster in tubes than in cages while others, such as <u>Prorocentrum minimum</u>, did not grow in tubes (Table 3). Similar results were obtained for other trials.

Initial biomass levels for the three trials were high and ranged from 13 to 16 µg Chl_a l^{-1} with final concentrations in the range of 40 to 90 µg Chl_a l^{-1}. Nutrient availability may have limited the final yield and growth rate in both the tubes and cages. Since growth in diffusion cultures is a function of the biomass present and the nutrient flux rate then reducing the biomass should increase nutrient availability and yield increased growth rates if nutrient transport limited growth. Conversely, growth rates in dialysis tubes

TABLE 2: A comparison of growth in dialysis tubes and plexiglass - Nuclepore $^{(R)}$ filter cages. Growth estimates (±1 S.D.) based on changes in chlorophyll$_a$ concentration.

Trial	Sample	Dilution	μ day^{-1}
I	Tube 1	0	0.29 ± 0.06
	Cage 1	0	0.30 ± 0.10
	Cage 2	0	0.07 ± 0.07**
II	Tube 1	0	0.83 ± 0
	Tube 2	0	0.72 ± 0.03
	Cage 1	0	0.84 ± 0.02
	Cage 2	0	0.86 ± 0.01
III	Tube 1	50%	1.02 ± 0.15
	Tube 2	50%	0.87 ± 0.04
	Cage 1	50%	0.61 ± 0.06*
	Cage 2	50%(1)	1.23 ± 0.01*

** - significant difference at P = .01
 * - significant difference at P = .05
(1) - Cage 2 light reduced by an additional 50%

should be equivalent in diluted and undiluted cultures if nutrient transport rates were not limiting.

In 4 of 5 trials made in Tampa Bay and in one trial from the oligotrophic central Gulf of Mexico (Trial 6) dilution of the phytoplankton standing crop lead to an increase in the average growth rates of that assemblage (Table 4). However, significant differences, were found in only 5 of 11 sample pairs. Thus high biomass and nutrient transport may have limited the growth rate of both the Tampa Bay and Central Gulf populations. These results also suggest that the populations were nutrient limited *in situ* in the sense that input and/or regeneration processes were supporting growth rates and a standing crop lower than the potential maximum. Dilution of the original population reduced the biomass to below the transition point of the exponential and linear growth phases of the culture when membrane transport limits growth. Thus, the population was able to grow exponentially at a rate set by environmental variables rather than by nutrient transport rates through the membrane.

TABLE 3: Growth rates of selected species from populations incubated in dialysis tubes and plexiglass - Nuclepore(R) filter cages.

	K (doublings day^{-1})	
Species	Tube	Cage
Skeletonema costatum	1.40	0.95
Nitzschia pungens	1.91	0.72
Nitzschia closterium	2.72	2.08
Leptocylindrus minimus	0.74	1.04
Chaetoceros compressus	1.16	0.55
Peridinium quinquicorne	1.25	1.60
Prorocentrum minimum	0.00	0.92

    The impact of grazing on both diluted and undiluted caged natural populations must also be considered. Landry and Hasset (1982) used the apparent increment in growth rate obtained from a sequential dilution of natural populations as an estimate of community grazing pressure by microzooplankton. Their hypothesis assumes that grazers will be diluted to a greater extent than the phytoplankton population. This lowers the total grazing impact. They must also assume that phytoplankton growth rates are identical at all dilutions as long a nutrient levels remain constant during the incubation period. Grazing is calculated as the slope of the regression line for growth vs. dilution. The Y-intercept yields an estimate of the apparent μ_{max}. In one of their experiments maximum growth rates were comparable for populations incubated in bottles with added nutrients and in dialysis tubes incubated in situ. In two other experiments the addition or deletion of nutrients to bottle enclosed populations yielded comparable growth rates at high dilution levels whereas depressed growth rates were found at no or low dilutions without nutrient additions. The authors state that this latter response "leads to an erroneous impression of the magnitude of the micro-zooplankton grazing impact". I feel that their data demonstrates that nutrient limitation had a greater impact on the growth of the population than did grazing, especially since populations at high dilutions had comparable growth rates with or without nutrient additions. Both of our results suggest that the dilution method yields elevated growth rates and would be useful as an indication of nutrient limitation in a natural population and to obtain an estimate of the carrying capacity for a given set of environmental variables.

TABLE 4: Growth rates (μ day^{-1}) of natural populations from Tampa Bay (Runs I-V) and the central Gulf of Mexico (Run VI) based on changes in chlorophyll concentrations. Values are ±1 S.D. Growth rates in the diluted samples were compared to the undiluted sample with ANOVA and a t-statistic.

Trial	Sample, % dilution	μ day^{-1}
I	0	0.29 ± 0.06
	20	0.36 ± 0.09*
	50	0.33 ± 0.09
	70	0.72 ± 0.23**
	90	0.49 ± 0.05**
II	0	0.77 ± 0.07
	50	0.94 ± 0.12*
III	0	0.45 ± 0.16
	50	0.18 ± 0.14**
IV	0	0.74 ± 0.21
	25	0.95 ± 0.08*
	50	0.80 ± 0.12
V	0	0.53 ± 0.47
	25	0.64 ± 0.36
	50	0.82 ± 0.21
VI	0	0.68 ± 0.31
	50	1.61 ± 0.23*

** - significant difference at P = 0.01
* - significant difference at P = 0.05

Examples of growth rates measured during short-term incubation of undiluted natural populations in dialysis tubes are presented in Table 5. Several diatom species in the oligotrophic slope waters of the

TABLE 5: Growth rates (doublings d^{-1}) of numerically abundant diatoms and the total population incubated in dialysis tubing during cruise B-8123. When two values are given they are for replicate tubes.

			K - Doublings Day $^{-1}$	
Station	Location	Species	count	Chl
4	shelf break	H. hauckii	4.37	1.76
		R. alata	4.09	0.92
		R. alata f. gracillissima	4.53	
			4.37	
		N. pungens	2.94	
		Total population	4.20	1.76
			3.84	0.92
7	slope	F. hyalina	4.41	1.70
			2.46	0.62
		Total population	4.26	1.70
			2.52	0.62
12	shelf	N. longissma	1.39	0.79
			1.59	0.51
		N. pungens	0.11	
			0.29	
		R. alata	0.32	
			0.41	
		R. robusta	1.51	
		R. setigera	0.57	
		S. unipunctata	1.84	
		Total population	0.79	0.79
			0.91	0.51
13	nearshore	R. alata	0.22	
		Total population	0.23	0
17	nearshore	R. alata	1.52	0.35
		R. alata f. gracillissima	0.91	0.53
		R. fragilissima	1.38	
		R. setigera	1.11	
		R. stolterforthii	1.14	
		C. lorenzianum	0.22	
		Total population	1.09	0.35
				0.53

eastern Gulf of Mexico had doubling rates in excess of 4 day^{-1}. Nutrient flux could support this growth rate. The ammonium concentration at stations 4 and 7 was approximately 0.5 µg at l^{-1} with nitrate undetectable. Using my values for ammonium transport, the flux rate could support 4.4 divisions in a 24 hour period at the initial biomass levels of 0.07 and 0.05 µg Chl$_a$ l^{-1} for stations 4 and 7 respectively. Thus the population at both stations was growing at the maximum rate diffusion would allow. Growth rates of several species at other stations were not unexpectedly high.

There was also considerable areal variation in the growth rate of the same species. The growth rate of <u>Rhizosolenia alata</u> ranged from 0.22 to 4.09 divisions day^{-1} (Table 5). Such differences may be attributed to variations in biomass between stations. Ammonia and nitrate concentrations were essentially the same at all stations (about 0.5 µg at l^{-1}). However, the initial biomass of each culture varied by more than an order of magnitude from less than 0.1 µg Chl$_a$ l^{-1} for the shelf break and slope station to greater than 1 µg Chl$_a$ l^{-1} at near-shore stations. Therefore, flux rates may have limited growth at the inshore stations.

Growth rates based on chlorophyll consistently underestimate those calculated from cell counts. This suggests that a lag period occurs with respect to growth as measured by chlorophyll and/or that the populations were adapted to a lower light regime. The results also demonstrate that not all species within an assemblage grow equally fast. Other species in each of the assemblages in Table 5 exhibited no growth or a decline. Lack of growth in cage cultures may reflect the inablility of some species to tolerate the turbulant conditions inside the tubes (Furnas, 1982a,b; Sakshaug and Jensen, 1978; Vargo, 1976) or represent the true situation <u>in situ</u>; i.e. some species exhibit exponential growth and outcompete others while some species maintain a minimal population level or decline.

General Discussion

What we are really looking for is the actual growth or production rates of natural populations unmodified by cage "bars". Do filter-walled cages or dialysis tubes provide the best estimate or will division rates based on microscopic observation of paired cells, such as those reported by Swift and Durbin (1972), Weiler and Karl (1979), Weiler and Chisholm (1976) and Weiler (1980), be the only way to get true, in situ growth rates? Some growth rates of natural populations enclosed in dialysis tubes are high relative to other published data (Eppley, 1972; 1981). Growth rates in cages exceed these estimates. If my results (Table 5) reflect the actual potential maximum exponential growth rates for several diatom species, then additional consideration must be given to earlier in situ estimates. If division rates for natural populations of S. costatum in cage cultures are in the range of 4.8 - 5.9 day^{-1} (Furnas, 1982a,b) then we must also reconsider some laboratory results. Such growth rates are rarely attained in culture under supposedly ideal growth conditions.

Differences observed between growth rates measured using dialysis tubes and cages should not be viewed as entirely being a methodological problem. Growth rates based on both techniques are high, therefore we may have to reconsider some concepts of the potential maximum growth rates that can be achieved by diatoms in situ. How common are growth rates of greater than 3-4 doublings day^{-1} in diatoms? Are they an artifact of the method in the sense that since cages act as nutrient concentrators, encapsulating a population simply allows it to grow under nutrient sufficient and otherwise ideal conditions thereby achieving a true μ_{max}?

The hypothesis presented by Goldman et al. (1979) and Goldman (1980) states that phytoplankton populations in oligotrophic, oceanic waters are not nutrient limited and have high relative growth rates ($\mu/\mu_{max} \cong \mu_{max}$). A low, steady-state biomass must be maintained by a combination of nutrient regeneration and loss rates.

Growth rates for several species in the slope and shelf-break waters (Table 5) tend to support their hypothesis, however, not all species grew at such high rates. It would not be illogical to assume that some species in an assemblage would have the intrinsic ability to outcompete others over short time periods. Certainly the high growth rates of several estuarine and coastal diatoms presented by Furnas (1982a,b) and my results (Table 5) indicate that some diatoms do have intrinsically high growth rates. There is no reason to assume that rapid growth would not be realized in natural populations if environmental conditions can support high turnover rates. Temperature and light are not normally considered to be limiting growth in oceanic surface waters. My dilution series results (Table 4) suggests that the Central Gulf population was nutrient limited. Any type of a cage culture acts as a nutrient concentrator. Enclosing a population in a cage sets up a high diffusive gradient between the inside and outside of the cage. In effect nutrient availability increases until membrane transport equals uptake. Populations enclosed within the cage should therefore grow at their intrinsic maximum rate until biomass increases to the point when competition for available nutrients and membrane transport again limits growth. Thus, cage cultures of natural populations, when combined with a dilution series might be used to obtain an estimate of the potential μ_{max} for species within an assemblage.

Obviously questions about the differences in growth rates observed in cages and tubes must be resolved. The relationship between dilution and elevated growth rate leads to intriguing questions about nutrient limitation and growth rates in situ and the relationship between biomass and growth rate. Is this a response to nutrient availability or to grazing pressure?

Cage culture techniques with natural populations incubated for relatively short time frames is an effective method for determining the relative and/or potential growth and production rates of individual species comprising an assemblage. After all it is both biomass and growth rate which determines the production in an area. The importance of the growth of one species relative to another in the scheme of trophic interaction must, at some point, also be included in the equation.

Literature Cited

Carpenter, E.J. and J.S. Lively (1980) Review of estimates to algal growth using ^{14}C tracer techniques. In: Falkowski, P.G. (ed.) Primary Production in the Sea. Brookhaven Symposium on Biology No. 31. Plenum Press, N.Y., p.161.

Chisholm, S.W., and J.C. Costello (1980) Influence of environmental factors and population composition on the timing of cell division in Thalassiosira fluviatilis (Bacillariophyceae) grown on light/dark cycles. J. Phycol. 16: 375-383.

Chisholm, S.W. (1981) Temporal Patterns of Cell Division in Unicellular Algae. In: Platt, T.(ed.), Physiological Bases of Phytoplankton Ecology. Canadian Bulletin of Fisheries and Aquatic Sciences No. 210. Department of Fisheries and Oceans, Ottawa, p. 150.

Eppley, R.W. (1972) Temperature and phytoplankton growth in the sea. Fish. Bull. (U.S.) 70: 1063-1085.

Eppley, R.W. (1980) Estimating phytoplankton growth rates in the central oligotrophic oceans. In: Falkowski, P. (ed.), Primary Productivity in the Sea. Brookhaven Symposia in Biology No. 31, Plenum Press, N.Y., p. 231.

Eppley, R.W. (1981) Relations between nutrient assimilation and growth in phytoplankton with a brief review of estimates of growth rate in the ocean. In: Platt, T. (ed.), Physiological Bases of Phytoplankton Ecology. Canadian Bulletin of Fisheries and Aquatic Sciences No. 210. Department of Fisheries and Oceans, Ottawa, p. 251.

Eppley, R.W., F.M. Reid, J.D.H. Strickland (1970) The ecology of the plankton off La Jolla, California, in the period April through September 1967 Strickland J.D.H. (ed.), pt. III. Estimates of phytoplankton crop size, growth rate and primary production. Bull. Scripps Inst. Oceanogr. 17: 33-42.

Fitzwater, S.E., G.A. Knauer and J. H. Martin (1982). Metal contamination and its effect on primary production measurements. Limnol. Oceanogr. 27: 544-551.

Furnas, M.J. (1982a) An evaluation of two diffusion culture techniques for estimating phytoplankton growth rates *in situ*. Mar. Biol. 70: 63-72.

Furnas, M.J. (1982b) Growth rates of summer nanoplankton (<10μm) populations in lower Narragansett Bay, Rhode Island, U.S.A.. Mar. Biol. 70: 105-115.

Gieskes, W.W.C., G.W. Kraay and M.A. Baars (1979). Current ^{14}C methods for measuring primary production: Gross underestimates in oceanic waters. Neth. J. Sea Res. 13(1): 58-78.

Goldman, J.C., J.J. McCarthy and D.G. Peavy (1979) Growth rate influence on the chemical composition of phytoplankton in oceanic waters. Nature, Lond. 279: 210-215.

Goldman, J. C. (1980) Physiological processes, nutrient availablilty, and the concept of relative growth rate in marine phytoplankton ecology. In: Falkowski P.(ed.), Primary Productivity in the Sea. Brookhaven Symposia in Biology No. 31. Plenum Press, N.Y., p.179.

Holm-Hansen, O., C.J. Lorenzen, R.W. Holmes and J.D.H. Strickland (1965). Fluorometric determination of chlorophyll. J. Cons. Perm. Int. Explor. Mer 30: 3-15.

Jensen, A., B. Rystad and L. Skoglund (1972) The use of dialysis cultures in phytoplankton studies. J.exp. mar. Biol. Ecol. 8: 241-248.

Jensen, A. and B. Rystad (1973) Semi-continuous monitoring of the capacity of sea water for supporting growth of phytoplankton. J. exp. mar. Biol. Ecol. 11: 275-285.

Landry, M.R. and R.P. Hassett (1982) Estimating the grazing impact of marine microzooplankton. Mar. Biol. 67(3): 283-288.

Nelson, D.M. and L.E. Brand (1979) Cell division periodicity in 13 species of marine phytoplankton on a light/dark cycle. J. Phycol. 15: 67-75.

Owens, O. v-H., P. Dresler, C.C. Crawford, M.A. Taylor and H.H. Seliger (1977) Phytoplankton cages for the measurement *in situ* of the growth rates of mixed natural populations. Chesapeake Sci. 18: 325-333.

Sakshaug, E. (1977) Limiting nutrients and maximum growth rates for diatoms in Narragansett Bay. J. exp. mar. Biol. Ecol. 28: 109-123.

Sakshaug, E. and A. Jensen (1978) The use of cage cultures in studies of the biochemisty and ecology of marine phytoplankton. Oceanogr. Mar. Biol. Ann. Rev. 16: 81-106.

Sakshaug, E. (1980). Problems in the methodology of studying phytoplankton. In: Morris, I. (ed.) The Physiological Ecology of Phytoplankton. Univ. of California Press p. 57.

Schultz, J.S. and P. Gerhardt (1969) Dialysis culture of microorganisms: design, theory and results. Bact. Rev. 33: 1-47.

Smith, R.C. and K.S. Baker (1980) Biologically effective dose transmitted by culture bottles in ^{14}C productivity experiments. Limnol. Oceanogr. 25(2): 364-366.

Swift, E. and E.G. Durbin (1972) The phased division and cytological characteristics of Pyrocystis spp. can be used to estimate doubling times of their populations in the sea. Deep-Sea Res. 19: 189-198.

Vargo, G.A. (1976) The influence of grazing and nutrient excretion by zooplankton on the growth and production of the marine diatom Skeletonema costatum (Grev.) Cleve, in Narragansett Bay, 126pp. Ph.D. Thesis, Univ. of Rhode Island, Kingston, R.I..

Vargo, G.A. (1979) The contribution of ammonia excreted by zooplankton to phytoplankton production in Narragansett Bay. J. Plankton Res. 1: 75-84.

Weiler, C.S. (1980) Population structure and in situ division rates of Ceratium in oligotrophic waters of the North Pacific central gyre. Limnol. Oceanogr. 25: 610-619.

Weiler, C.S. and S.W. Chisholm (1976) Phased cell division in natural populations of marine dinoflagellates from shipboard cultures. J. exp mar. Biol. Ecol. 25: 239-247.

Weiler, C.S. and D.M. Karl (1979) Diel changes in phased-dividing cultures of Ceratium furca (Dinophyceae): nucleotide triphosphates, adenylate energy charge, cell carbon and patterns of vertical migration. J. Phycol. 15: 384-391.

OBSERVED CHANGES IN SPECTRAL SIGNATURES OF NATURAL PHYTOPLANKTON POPULATIONS : THE INFLUENCE OF NUTRIENT AVAILABILITY

Ch.S. YENTSCH and A. PHINNEY

Bigelow Laboratory for Ocean Sciences, W. Boothbay Harbor, Maine, 04575 USA.

INTRODUCTION

A question often asked - "Is the high species diversity one observes in phytoplankton populations the result of differences in the physiology of specific species?" Perhaps classical taxonomy is too "fine-tuned" to identify environmental factors which affect individual species' physiology.... It is suggested that the environmental physiologist might be better served by measurements of features that characterize the basic energy flow and these be used as a taxonomic index. One means of doing this involves the measurement of spectral characteristics of cellular fluorescence from different types of photosynthetic autotrophs (Yentsch and Yentsch, 1979).

Historically, early studies of plant physiology established the potential of utilizing fluorescent signatures. The pioneering research of Engleman in the 1860s recognized that plants and algae through diverse pigmentation, could absorb light of different colors and utilize it for photosynthesis. As more physiological evidence accumulated, it became apparent that different morphological and hence, taxonomic groups, possessed different suites of pigments. This classification is based on color and each color group has a distinctive color fluorescent signature.

DIFFERENTIATION OF COLOR GROUPS

Taking the bulk of all known photosynthetic autotrophs in the ocean, there appears to be at least three ways in which light is harvested. These can be differentiated by the light absorbed by photosynthetic pigments which are accessory to the main photosynthetic pigment, chlorophyll a. Individual color groups of photosynthetic autotrophs can be indexed using the different types of pigments accessory to chlorophyll a. These are, chlorophylls b and c, carotenoid proteins (this refers to a pigment-protein complex), such as fucoxanthin and peridinin, and the biliproteins, phycoerythrin and phycocyanin (see Fig. 1).

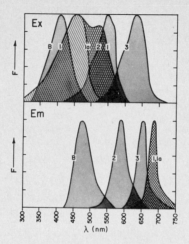

Figure 1. Fluorescence spectral signatures (F = fluorescence, E = excitation) for different algal color groups. Scheme 1, diatoms and dinoflagellates; scheme 1a, green algae; scheme 2, cryptomonads and cyanobacteria. B is the spectral signature for the bioluminescent substance, luciferin. From Yentsch and Phinney, 1980).

Table 1 shows the chlorophyll accessory pigment relationship common to organisms in the sea. In the right hand column of this table, is a ratio ($E_{530}:E_{450}$) which reflects the efficiency at which chlorophyll a fluoresces excited by either of these two wavelengths. For example, a large ratio, 0.8:0.9 indicates that the two wavelengths induce chlorophyll fluorescence with almost equal efficiency, whereas a low ratio indicates that 450 nm is more effective than 530 nm in inducing chlorophyll fluorescence. These ratios are indicative of the degree which light is absorbed by the accessory pigments and transmitted as chlorophyll fluorescence. The high ratios are observed in diatoms and dinoflagellates are due to the fact that they contain considerable quantities of carotenoid proteins. Note in the table that coccolithophores have lesser amounts of this type of pigment and green algae contain little or no carotenoid proteins. In the case of the organisms with chromoproteins, the principle phycobilin pigment in natural marine populations is phycoerythrin (Yentsch and Phinney, 1984). In this case, it is the phycoerythrin that is providing the major light absorption between the spectral regions of 532 and 550 nm. In the case of the cyanobacteria, the ratio ($E_{530}:E_{450}$) is negligible since the transfer of energy from phycoerythrin to chlorophyll is undetectable (Yentsch and Phinney, 1984).

Table 1 shows the possible spectral interpretations for marine species using two excitation wavelengths, measuring two wavelengths of emission. All algae, with the exception of the cyanobacteria,

Table 1.
Chlorophyll-accessory pigment relationships.

F = fluorescence by chl a at 680 nm
E = wavelength of excitation for chl a fluorescence

Organism	Primary accessory pigments	Chl a F ($E_{530}:E_{450}$)
Dinoflagellates	Carotenoids, Chl c	0.8 - 0.9
Diatoms	Carotenoids, Chl c	0.7 - 0.8
Coccolithophores	Carotenoids, Chl c	0.3 - 0.4
Green flagellates	Chl b	0.1 - 0.2
Cryptomonads	Phycobilins	0.7
Cyanobacteria	Phycobilins	-

will emit fluorescence when excited by light at 450 nm. Organisms with carotenoid proteins, such as diatoms and dinoflagellates, will fluoresce at 680 nm. Cryptomonads will also fluoresce at 680 nm when excited by wavelengths of 530 nm, however, in this case, the accessory pigmentation is the phycobilin, phycoerythrin and not a carotenoid protein. Both cyanobacteria and cryptomonads will fluoesce at 580 nm, which is the primary band for phycoerythrin fluorescence, when excited by 450 or 530 nm.

FLUORESCENCE SIGNATURE EXTREMES IN OPEN OCEANIC POPULATIONS

For the past five years, we have measured spectral fluorescence in particulate matter collected from the world's oceans. Since the number of fluorescent cells in the particulate matter is generally low, we have had to concentrate particles on membrane filters (0.45 μ) and mount these in the light path of the conventional spectrofluorometer. The details of this method can be found in Yentsch and Phinney (1984). The use of membrane filters in this manner also alleviates spurious signals that occur from cells settling or being moved about in cuvettes.

The extremes in spectral signatures that we have observed are shown in Fig. 2. These extremes illustrate the spectral differences observed in populations taken in eutrophic waters as opposed to those populations taken in oligotrophic waters. In these two examples, the

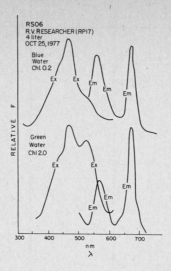

Figure 2. Fluorescent spectral extremes across a color front. Gulf of Mexico, October, 1977. Ex = excitation, Em = emission. (From Yentsch and Phinney, 1984).

spectral signatures were obtained from waters on either side of a strong color front, which was formed by nutrient rich Mississippi water flowing into blue, oligotrophic waters of the Gulf of Mexico. To distinguish these two water masses, we have designated the spectral signatures as "green" and "blue" water. Note that the concentration of chlorophyll in the "green" water is 10 times greater than that of "blue" water. The major difference in the two excitation spectra confirms the effectiveness at which light is absorbed at 530 nm. In the "green" water population, this wavelength is absorbed much more effectively than the "blue" water population; the distinguishing feature of the excitation spectra is the strong shoulder at 530 nm. This shoulder is reduced in the blue water population. These differences are interpreted to be due to the presence of organisms having large concentrations of carotenoid proteins as accessory pigments which, in turn, suggests the occurrence of diatoms and dinoflagellates in this water mass. This suggestion was confirmed by microscopic observations in water samples taken from the two water types. The "green" water population is made up largely of diatoms and dinoflagellates while very few species were found in the "blue" water. Both water types contained organisms with a considerable amount of phycoerythrin. However, if the phycoerythrin level is considered in relation to the chlorophyll emission, there is considerably more phycoerythrin fluorescence in the "blue" water population than in the "green" water.

CHANGES IN FLUORESCENT SIGNATURES OF NATURAL POPULATIONS ASSOCIATED WITH NUTRIENT ENRICHMENT

We have accumulated evidence that suggests that the major change in fluorescent signatures is associated with the transition from oligotrophic to eutrophic waters. One of the best examples of these changes can be found in the frontal regions which separates stratified from unstratified waters, namely in a region of tidal mixing. Figures 3-5 show properties of water masses adjacent to and off Georges Bank. The extremes in the vertical mixing in this section can be observed by examination of the slope of the isopycnal surfaces (Fig. 3). Approaching Georges Bank, the isotherms change from

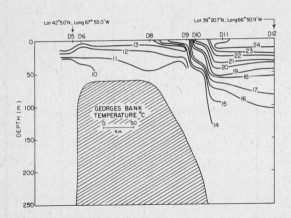

Figure 3. Temperature section between Bermuda and the Gulf of Maine. July 22 - August 8, 1980.

horizontal to almost vertical, reflecting the effects of mixing by tidal currents and bottom friction. A change in the thermal structure is reflected in the distribution of nutrient rich water in this section (Fig. 4). In the region of tidal mixing, there is an increase in nitrogen rich water and increased chlorophyll concentrations (Fig. 5). Associated with chlorophyll-rich and chlorophyll-poor waters are marked changes in the fluorescent parameters (Figs. 5-7). In regions of high chlorophyll concentration (greater than 0.5), the $E_{530}:E_{450}$ ratio (Fig. 6) ranges between 0.5 and 0.8. In

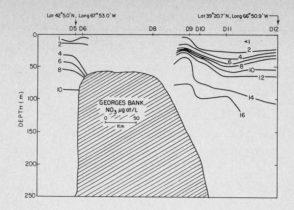

Figure 4. Nitrate-nitrogen section between Bermuda and the Gulf of Maine. July 22 - August 8, 1980.

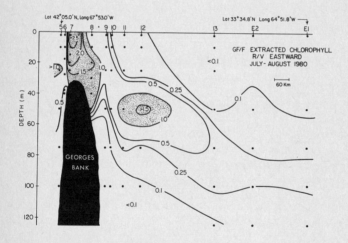

Figure 5. Extracted chlorophyll section between Bermuda and the Gulf of Maine. July 22 - August 8, 1980.

regions where chlorophyll concentrations are low (less than 0.5), the ratio ranges between 0.3 and 0.5. The contours of phycoerythrin fluorescence (Fig. 7) strongly resemble the contours of the $E_{530}:E_{450}$ ratio. The highest values for both parameters occurred adjacent to the Bank, that is on the seaward side with the maximum concentration with depth at approximately 50 m.

Along with the measurement of fluorescent parameters, direct cell counts of fluorescent particles were made at selected stations

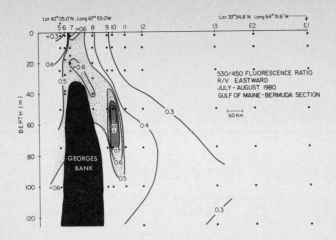

Figure 6. Section of $E_{530}:E_{450}$ fluorescence ratio between Bermuda and the Gulf of Maine. July 22 - August 8, 1980.

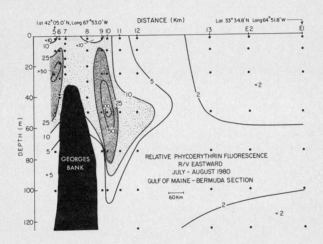

Figure 7. Relative phycoerythrin fluorescence section between Bermuda and the Gulf of Maine. JUuly 22 - August 8, 1980.

in the surface waters using epifluorescent microscopy. Procaryote cells (orange fluorescence) exceeded eucaryote cells (red fluorescence) by 100 times the oligotrophic waters. However, in the eutrophic waters near and on Georges Bank, procaryotes exceeded eucaryotes by only ten times, indicating the dominance of a population on and around Georges Bank of largely diatoms and dinoflagellates. These groups of organisms were virtually absent in the oligotrophic waters seaward of the Bank.

While the above data represents an example of spectral changes in a horizontal plane, these changes are also accentuated in vertical profiles (Fig. 8). The chlorophyll maximum occurs within the thermocline and coincides with the top of the nitrate/nitrogen nutracline (Fig. 8). The chlorophyll values above and below this maxima are about ten times less than values within the chlorophyll maximum. The $E_{530}:E_{450}$ ratio is mirrored by the chlorophyll and phycoerythrin fluorescence. These profiles suggest that, for the total population of the water column, procaryotes and eucaryotes have their largest accumulation within the chlorophyll maximum. Groups of organisms are less abundant above and below the maximum. Since we recognize the coincidence of chlorophyll maxima in the nitrate nutracline as a site of enrichment, one concludes that the spectral changes observed are associated with increasing abundance due to nutrient enrichment.

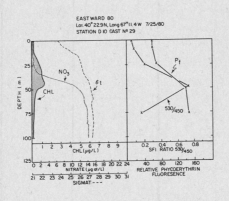

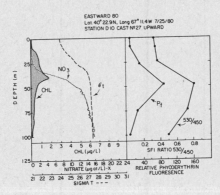

Figure 8. Selected parameters from two vertical profiles at station D10, July 25, 1980.

CONCEPT OF CHROMATIC ADAPTATION AS IT APPLIES TO SPECTRAL SIGNATURES IN NATURAL POPULATIONS OF PHYTOPLANKTON

Studies of accessory pigmentation of algae always seem to wind up at the doorsteps of the hypothesis of chromatic adaptation. Arguments for and against this hypothesis have been reviewed by Ramus (1981) and Dring (1982). Both workers have concluded that with regard to the vertical distribution of benthic algae in the sea, the light/pigment relationship seems to be compatible with variations in the levels of total irradiance which decreases with depth. Therefore, the wavelength dependency of the accessory pigmentation does not seem to be the important factor in the hypothesis.

Our fluorescence data on natural populations, i.e. the changes in spectra, cannot be interpreted in terms of light quality or in that fact, in terms of light intensity. Using the above observations, we have tried to establish the hypothesis that the relationship of spectral change is closely associated with changes in the abundance of algae, which, in the ocean, is regulated by the degrees of eutrophication or oligotrophy of the water masses. Therefore, the spectral fluorescent signatures are indicative of what one would call the "nutrient status" in natural populations. One can rightly ask if these are species assemblages or whether or not this chromatic change is due to cellular-induced synthesis or decomposition. With regard to this type of chromatic adaptation, procaryotes and eucaryotes are known to change their pigment composition with respect to light quality and quantity - we cannot discount that some of this is affecting the data that we have observed in these sections and profiles. However, direct microscopic counts indicated that these spectral changes are associated with the dominance of certain groups of organisms, such as diatoms and dinoflagellates and cryptomonads. Therefore, we conclude that the spectral signatures are largely affecting the assemblage of species and that these signatures, in a crude sense, reflect diversity shifts in natural populations and are associated with change in the nutrient level of the water masses.

POTENTIAL FOR REMOTE SENSING

In another paper (Yentsch and Phinney, 1984), we have attempted to point out the drawbacks in applying and interpreting fluorescent color group observations in terms of classical taxonomic classification; fluorescence signatures do not have enough information to adequately distinguish, for example, dinoflagellates from diatoms and/or coccolithophores. On the other hand, we have observed a close correspondence between the ratio $E_{530}:E_{450}$ and mean cell size of natural populations. Larger cell sizes in these populations (> 10μ) are associated with ratios of 0.6 to 0.8, whereas small cell sizes (< 10μ) have ratios of 0.3 or less. The significance of this finding is that fluorescence signatures appear to be able to provide information on the mean cell size of the populations which has the advantage that, with proper optical techniques, these estimates can be made remotely. Cell size of phytoplankton, that is surface to volume ratio relationships and how this is associated with physiology, is far from being understood, but we do know that the efficiency of nutrient uptake, respiration and photosynthesis are all apparent functions of cell size (Malone, 1980). We believe in this paper, that we have demonstrated that major spectral features of cellular fluorescence are associated with change in numbers and species of phytoplankton and these occur in the regions of transition between nutrient-rich and nutrient-poor waters. Such findings tend to support the hypothesis that a reduction of cell size is a requisite for phytoplankton survival in oligotrophic regions of the oceans. It is one of the more important goals in biological oceanography to delineate in time and space, these regions where marked transitions in population physiology occur. In looking at the enormous size of the ocean, this would appear to be an awesome task if it were not for the presence of remote sensing techniques.

Ocean color scanner and fluorescence instruments developed by NASA represent exciting new techniques for the biological oceanographer, and provide a means for assessment of large scale patchiness of the ocean in a manner only dreamed of by early practitioners in oceanography. The exciting aspect of the fluorescent LIDAR system is the ability to characterize the organisms present in the fashion of their spectral signatures. At the present time, LIDAR fluorescent systems are flown by aircraft at low altitudes. As mentioned, the instruments induce fluorescence by pulsing strong bursts of blue and green light into the water column. The present units are capable of resolving patches of a few meters in size, however, the coverage by aircraft is limited to the overall flight time of these units. One must recognize that the present instruments are merely prototypes and have their deficiencies, but that they are pioneers of an evolution that is just beginning.

SUMMARY

In this paper, the summarized results are:

1) the ratio $E_{530}:E_{450}$ is strongly correlated with chlorophyll abundance. We believe this reflects the influence of growth changes in nutrient limited conditions.

2) chlorophyll a fluorescence is much more effective excited by green wavelengths greater than 500 nm in coastal water populations than in populations of the open sea. We believe this is due to the fact that in coastal waters, accessibility to nutrients is much greater than in the oligotrophic waters.

3) fluorescence from biliproteins, such as phycoerythrin, is observed in all samples. The majority of this fluorescence is from small microorganisms such as cyanobacteria and/or cryptomonads (Yentsch and Phinney, 1984). For reasons that are unclear, these organisms seem to be able to persist in environments where growth tends to limit organisms with substantial quantities of carotenoid protein.

ACKNOWLEDGEMENTS

Most oceanographers recognize the interrelationship of biochemistry and field observations of biological oceanography. This interaction is vital to the science and its steady progression. The convenor of this symposium has championed this approach. Dr. Osmund Holm-Hansen's entire career has straddled the extremes of classical biochemistry to fundamental biological oceanography. Ozzie's role has been one of translating technique, philosophy, and interpretation for the biological oceanographic community. It is an activity that he has done extremely well and has received too few thanks. I wish to thank him for the invitation to this symposium and the very pleasant atmosphere he provided.

Support for this work was provided by NASA grant numbers NAGW-410 and NAS5-27742.

The is contribution number 83038 from the Bigelow Laboratory for Ocean Sciences. The authors acknowledge the able assistance of P. Boisvert in the preparation of this manuscript and J. Rollins in drafting the figures.

REFERENCES

Dring M. J. (1982) The Biology of Marine Plants. Willis A.J. and Sleigh M.A. (eds.) E. Arnold Publ. Ltd., London, 199 pp.

Malone T. C. (1980) Algal size. In : Morris I. (ed.) The Physiological Ecology of Phytoplankton. Blackwell, London, Vol. 7, pp. 433-463.

Ramus J. (1981) The capture and transduction of light energy. In : Lobban C. S. and Wynne M. J. (eds.) The Biology of Seaweeds. Blackwell Scientific Publications, Oxford, pp. 458-492.

Yentsch C. S. and Yentsch C. M. (1979) Fluorescence spectral signatures: the characterization of phytoplankton populations by the use of excitation and emission spectra. J. Mar. Res. 37 : 471-483.

Yentsch C. S. and Phinney D. A. (1980) Nature of particulate light in the sea. In : Nealson K. H. (ed.) Bioluminescence: Current Perspectives. Burgess Publ. Co., Minneapolis, pp. 82-88.

Yentsch C. S. and Phinney D. A. (1984) The use of fluorescence spectral signatures for studies of marine phytoplankton. In : Zirino A. (ed.) Chemical Oceanography, Advances in Chemistry, Series No. 208, American Chemical Society: Washington, D.C.

FLOW CYTOMETRY AND CELL SORTING : PROBLEMS AND PROMISES FOR
BIOLOGICAL OCEAN SCIENCE RESEARCH

C.M. YENTSCH, L.CUCCI and D.A.PHINNEY

The Jane MacIsaac Flow Cytometer/Sorter Facility
Bigelow Laboratory for Ocean Sciences
McKown Point, West Boothbay Harbor, Maine 04575, USA.

INTRODUCTION

With the sole exception of the measurement of phytoplankton fluorescence, biological oceanographers live within the constraint of having few methods which provide real time continuous data. Most methods are discrete measurements of the "whole" community.

Autoanalyzers, electrodes and Coulter volume particle analyzers are examples of instruments borrowed from biomedical laboratories for use in ocean sciences. We believe that the flow cytometer/sorter (Horan and Wheeless, 1977; Kruth, 1982; Melamed et al., 1979; Trask et al., 1982) is another tool which has great potential for ocean science research. We are just now in the initial phases of protocol development and focus, with the eventual goal being to have such instrumentation as an effective real time unit as a shipboard operation. In this manuscript, we discuss both the promises and problems with this new methodology.

FLOW CYTOMETRY AND SORTING: INDIVIDUALS MAKE THE DIFFERENCE

The flow cytometer/sorter can make simultaneous measurement of multiple properties of individual cells and particles at a rapid rate. Why is this important? Let us take the analogy of a sports stadium. With a quick glance, it is easy to estimate whether the stadium is full or empty; this information is of limited usefulness. If we want to subdivide the crowd to know how many are fans of team A and how many are fans of team B, and if every person displayed team colors, we could get a relatively good breakdown by eyeballing the crowd. But instead, we want to know several properties or parameters, such as team affiliation, height and weight, sex and age of the individual. These data are only obtainable if we make measurements not of the crowd, but of each individual as they file past an observation point. The flow cytometer measures various characteristics of cells/particles as they exit single file from the flow chamber (Fig. 1). On the basis of specific criteria we can then direct the cells to sort right or sort left.

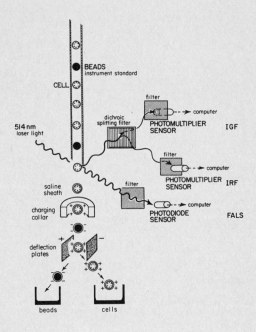

Figure 1. Fate of cells and standard beads as they pass through the flow cytometer with the sort criteria set to separate cells from beads based on a difference in wavelength or intensity of fluorescence emission. IGF = integrated green fluorescence; IRF = integrated red fluorescence; FALS = forward angle light scatter (1.5-19°). Not shown is the droplet breakoff point which is immediately below the point of analysis.

Why this development is so significant is that we are now able to exploit signals to quantify and define variance, and separate subpopulations of cells and particles in the 1.0 to 150 μm size range. Before, we were only able to make measurements on all of the organisms of interest in a given volume of seawater. The result was an "average" value. Now, each individual cell is analyzed. Returning to the sports stadium analogy, this would mean that characteristics of every individual in a stadium of 100,000 could be "fingerprinted" in less than 2 minutes!

Conveniently, phytoplankton have their own innate fluorescing material, the light-harvesting pigments of chlorophyll and accessory pigments (Yentsch and Yentsch, 1979). With flow cytometry, high sensitivity is attained using laser excitation of both pigments (termed autofluorescence) and applied stains (termed induced fluorescence) or fluorescent-conjugated antibodies (termed immunofluorescence). Soon it will be feasible to use chlorophyll fluorescence to distinguish phytoplankton, protein stains to detect bacteria and small animals, and a have measurement of growth rate made directly using DNA stains. Also, light scatter can be a useful index for inorganic sediment particles as well as living and non-living organic particles (see Yentsch et al., 1983a).

WHAT IS FLOW CYTOMETRY/SORTING?

FLOW CYTOMETRY is a rapid analytical technique which measures properties of cells/particles in a stream flow. Combining optics, electronics and laser excitation, quantitative fluorescent signals derived from natural pigments, stains and/or antibodies form a "fingerprint" of multiple parameters from particles analyzed one-at-a-time.

CELL/PARTICLE SORTING is a technique which combines flow cytometry, computerized data acquisition and commands, and clever engineering. The saline sheath surrounding a cell/particle in the form of a droplet is electrostatically charged when preselected analysis criteria are met. The droplet containing the desired particle passes through deflection plates which diverts the droplet to either the left or to the right for collection.

HOW MUCH DOES THE INSTRUMENTATION COST?

There are a variety of instruments which can be fabricated from off-the-shelf equipment based on a mercury lamp epifluorescence microscope and a photomultiplier tube resulting in rapid measurement of one parameter (see Olson et al., 1983). Commercially available analysis units start at $50,000 USA which can measure multiple parameters, again using a mercury lamp light source. Fully versatile analysis units designed for multiple parameters with full sorting capabilities, dual laser light sources and multiple data acquisition and display systems, at the time of this writing run approximately

$170,000 USA. Additional options include an autocloning device, Coulter volume sensor, and extended data analysis systems.

WHAT WILL WE LEARN ABOUT THE OCEANS WHICH MERITS THIS APPROACH TO RESEARCH QUESTIONS?

Our specific interest is the phytoplankton -- microscopic plants which collect and convert solar energy into chemical energy of sugars and other compounds to form the first step of the food web. Collectively, phytoplankton are the pastures of the sea. Phytoplankton, however, are not the only particles in the 1 to 150 µm size range -- bacteria and microzooplankton occupy this same size range. Mud and silt particles and fine sands also overlap in size as does detritus, the decaying organic material present almost everywhere in the sea. While the complex size overlap once presented limitations on experimental approach and design, now, through fluorescence and light scatter characteristics we can exploit particle differences which can result in separation of subpopulations for biomass, elemental composition, cell physiology, enzymology, grazing studies, etc.

What we can say with assurance is that flow cytometry/sorting is a technological breakthrough which, for the first time, permits a close examination of individual cells in a rapid and statistically significant fashion. There are many important questions about the oceans which have captured scientists' curiosity for decades. Now it appears that we have a tool which should enable researchers to address questions previously intractable.

THE REQUIREMENTS FOR FLOW CYTOMETRY/SORTING

There are three basic requirements for flow cytometry/sorting. These are: first, the particles must be single cell/particle suspensions; second, at present, the particles must be within the size range of 1-150 µm; third, the particles should be in a concentration range of 10^5 to 10^6 cells per ml. Each of these requirements can present problems which must be dealt with beyond the instrumentation *per se* (Fig. 2).

TOWARDS OVERCOMING CURRENT LIMITATIONS

While phytoplankton cells commonly exist as single cells in a fluid medium, doublets as well as chain-forming diatoms and chain-forming cyanobacteria are common and must be reckoned with. Multicellular organisms present additional complications. Certainly

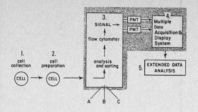

Figure 2. Aspects of flow cytometric analysis and sorting. Functions in stippled area (3,4) denote instrument capabilities. Important developmental needs include: 1) cell collection; 2) cell preparation/concentration/stain protocols; and 5) extended data analysis.

dissociation methods are vital to application of flow cytometry/sorting techniques. Development needs include single cell separations which preserve the integrity of the cell as a unit.

The current size range of the instrument's capability is between 1 to 150 μm which is almost ideal for phytoplankton research. The oceans have vast quantities of particles within this range. Still, by having size limits, we are subject to subsampling a community and must be aware of the resulting biases. This working range is restricted by the fact that a particle of 0.5 μm diameter is approaching the wavelength of light at the small end (e.g. 488 nm). Yet, some biomedical laboratories are dealing successfully with virus particles. To date, a 200 μm orifice controls the large end permitting measurement of particles up to 150 μm.

Cell concentrations in the marine environment vary substantially and are summarized crudely in Table I for oligotrophic, mesotrophic and eutrophic waters (Yentsch and Yentsch, in press). Note that with the exception of bacteria, cyanobacteria and other picoplankton, cell concentrations are far more dilute than the optimal range of 10^5 to 10^6 cells per ml. Thus a major developmental thrust must be for concentrating cells with minimal artifacts. At the moment in biomedical labs, cell elutriation looks promising - a process where counterflow during centrifugation permits concentration with only minor G forces applied. Success is being witnessed with oceanic particles as well (S. Pomponi, pers. comm.) Density gradient centrifugation and plankton Nitex netting can be effective, but inherent subsampling biases remain.

While samples can be run at ambient concentrations, one tends to bias interpretation based on the tremendous logging up of the smaller

Table I

Generalized table of cell concentrations on a per ml basis

	Oligotrophic open oceans	Mesotrophic shelf areas	Eutrophic coastal areas
bacteria .5-1.5 µ	10^6	10^7	10^8
cyanobacteria 1.0-1.5 µ	10^5	10^6	10^6
small autotrophs 3-10 µ	10^4	10^4	10^5
large autotrophs 10-50 µ	10^0	10^1	10^2
microzooplankton 10-150 µ	10^2	10^3	10^3
ciliates 50 µ	10^0	10^1	10^1

common cells. Remember that the instrument records fluorescence from each cell as "one event." Therefore, fluorescence from a small cyanobacterial cell is an event in the same manner that fluorescence from a large dinoflagellate or diatom cell is an event. Yet, the fluorescence intensities are recorded as substantially different. Thus from the data in Table I, one can assume that there will be 1,000 to 10,000 "events" from cyanobacteria (with their characteristic intensity per cell) recorded for every single event for a dinoflagellate or diatom (with their characteristic intensity per cell). An additional consideration here is that large cells (approximately 35 µm spheres) have up to 43,000 times the biomass as small cells (approximately 1 µm spheres). Yet, using the three decade log scale, we can make comparisons and relative measurements at the same time.

Sorting is accomplished by setting up a sort logic (up to 24 questions to which "yes" must be the proper response in each case). Two different sorts, one right and one left, can be accomplished simultaneously. Gates are easily set on fluorescence and/or light scatter criteria. The selection of these gates is critical to the experimental design. The cells are in a saline sheath. Individual droplets are broken off by vibration of a piezoelectric crystal. A time delay is calculated between analysis and sort mode. The droplet passes through a charging collar. A droplet containing a cell which meets pre-set criteria is charged, the rest are not. The droplet is then passed through deflection plates and sorted right (+) or left (-) depending upon the charge, or dropped into a center reject vial. While analysis can be accomplished with distilled water as the sheath fluid, the sheath fluid must be saline (at least 0.5%) to maintain charge for sorting. We customarily use filtered seawater (0.4 μm Nuclepore membrane) or appropriate culture medium for marine phytoplankton. However, the saline sheath would be expected to cause osmotic stress for freshwater aquatic cells. Even under conditions optimal for marine phytoplankton, the sheath fluid does substantially dilute the final recovered cells. In ^{14}C experiments, this has been estimated (by tracers) to be between 50-100 times dilution (P.J.leB. Williams, pers. comm.).

The instrument, designed with the objective of sort purity, has resulted in computer controlled anti-coincidence feedback loops which result in highly pure sorts to the right and to the left. The reject container, however, is in no way a "pure" sample. Contents will include 1) particles failing to meet sort criteria, 2) particles which meet sort criteria but are rejected due to coincidence, 3) debris, and 4) excess sheath fluid. If the objective is to recover most of the cells, the reject container must be resorted repeatedly, which further dilutes the cells with sheath fluid.

Due to computer speed limitations, effective sorting cannot be accomplished as rapidly as analysis. In general, while analysis can cope with up to 0.4 ml per minute, only 1 ml per hour can be sorted. Yield is dependent on particle concentration per ml and initial percentage of desired particles in the mixture/natural population. For example, one has to sort ten times as long to collect x number of cells if the desired cells are 2% of the population as compared with 20% of the population. For microscopic verification of sorted cells and electron microscope work, sufficient cell numbers are readily achieved. On the other hand, for subsequent physiological experiments or biochemical measurements, sort times are not always practical within a time frame permissible to insure that little or no change has occurred during the sort time itself.

We predict and urge a general thrust to develop micro-methods for elemental analysis, tracer incorporation, etc. which should result in the capability to deal with small numbers of cells (10^2-10^4 cells, irrespective of the fluid volume from which recovered). Such development would permit maximum utility of the flow cytometer/sorter and the necessary coupling with more traditional methods which must be evoked in order to assign real numbers to the relative axes which are the boundaries for flow cytometric measurements.

Coupled with a thrust to develop micro-methods, is the necessity to search for suitable reaction blockers for photosynthesis incubations, growth incubations, respiration incubations and enzyme activity experimentation. If the reactions are stopped adequately with little degradation of the cells and pigments, then appropriate sorting can be permitted to occur over hours vs. minutes. Sterile sorting is also possible and the instrument has a temperature controlled sampling vial holder which permits use of the vial as a reaction container.

One new dimension is certain to be the use of fluorescent stains as probes. Table II is a compilation of some of the stains commonly used in biomedical research. Stain and antibody development has evolved employing mammalian cells (Kruth, 1982), a cell environment which is spectrally relatively "competition-free." Phytoplankton cells, on the other hand, have pigments crowded into the spectrum. Chlorophyll a is common to green plants with fluorescence emission approximately 680 nm. In addition, various chromatic groups contain accessory pigments. In total, these accessory pigments and the chlorophyll pigments fill up the absorption regions ($\cong$ excitation wavelengths for fluorescence) in the visible spectrum. Note that the absorption of water competes with these plant pigments for light in the aquatic environment (Yentsch and Yentsch, 1983).

When vital staining is desired (a term used to denote that the cells remain viable and unaltered), stains must be carefully selected to avoid overlap of excitation and/or emission with algal pigments. Energy transfer is the phenomenon of concern. Basically, the light emitted by one fluorescent reaction can be reabsorbed by another molecule resulting in additional excitation and emission. A person unknowledgeable of this might attempt to measure emission in the spectral region where fluorescence is expected solely from one molecule and in fact the emission occurs in another spectral region. In doing so, the fluorescence emission from the first molecule is "less" than expected. The molecules do not need to touch each other for this transfer to take place, merely the light emitted by one molecule excites the second molecule.

Another option is to fix the algal cells and extract the pigments in an organic solvent (such as 50% ethanol or methanol) as has been used effectively by Olson et al. (1983). Yentsch and

Table II. Fluorescence excitation and emission of a sampling of fluorescent probes.

	"best" laser line for excitation	emission maximum
DNA*		
Hoechst dye family	365	460
DNA*		
DAPI	365	460
DNA		
mithramycin	488	560
DNA + RNA		
propidium iodide	488, 514	625
DNA + RNA*		
acridine orange	488	530
MEMBRANE POTENTIAL/LIPIDS*		
cyanine dye family	488	530
Di-O-C_7		
PROTEINS		
**fluorescein isothiocyanate (FITC)	488	517
**rhodamine isothiocyanate (RITC)	514	590
fluorescamine	365	490
CELL VIABILITY		
fluorescein diacetate (FDA)	488	550 or 630

* "vital" stains
** common fluorescent tags for antibodies

co-workers (1983) opted to photo-oxidize the pigments and then proceed with staining protocols.

METHODS: EXAMPLES OF SOME SUCCESSFUL APPLICATIONS OF FLOW CYTOMETRY/SORTING

In summary, we would like to present some data to demonstrate some of the problems and the promises elaborated. The first essential is a careful selection of standards and controls. Gains, laser power, gates, etc., are all arbitrarily set on the instrument. Therefore, for the data presented here, we have used an instrument standard (polyvinyl fluorescent microspheres 10 μm in diameter) and fine-tuned the variables until a 2.0% or less C/V (coefficient of variation) of light scatter and fluorescence was achieved. At that point we accept that the instrument is in a starting position -- instrument drift throughout the time of the experiment is checked by adding these beads as an internal standard, or monitoring position (channel) and C/V both before and after experimentation.

A. Experimentation with clonal cultures:

Many cultures are ideal candidates for flow cytometric analysis. Dinoflagellates, however, grow to cell densities which are more dilute than optimum. Cultures of the New England red tide toxic dinoflagellate *Gonyaulax tamarensis* var. *excavata* (clone GT-429) were grown in F/2 medium at 15°C under various photon flux densities in continuous light for 11 days. 100% represents ~ 266 $\mu Ein \cdot cm^{-2} \cdot s^{-1}$. Flow cytometric analysis of 2000 cells (following centrifugation at 2000 G for 5 m to concentrate cells) indicated an increase in chlorophyll fluorescence emission (>630 nm; log scale LIRFL) and parallel decrease in cell size as measured by forward angle light scatter (FALS, linear scale) (see Fig. 3). Numbers of cells analyzed is sufficient to permit rigorous statistical analysis. It can be quickly demonstrated that there are dramatic 1) increases in cell chlorophyll fluorescence emission; 2) decreases in cell size; and 3) decreases in coefficient of variation (C/V) of cells at reduced photon flux densities. Corresponding division rates for cells were 0.36 day^{-1} at 100%; 0.45 day^{-1} at 50%; and 0.20 day^{-1} at 25%.

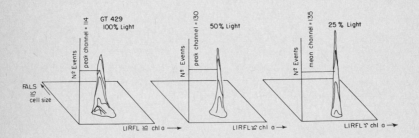

Figure 3. Three-dimensional bivariate 64 channel x 64 channel plots of fluorescence on a three decade log scale (LIRFL) $\cong$ chl *a* and forward angle light scatter (FALS) $\cong$ chl size plotted against relative number of events. 2000 cells were analyzed in each case using 50 mW 488 nm laser line on a Coulter EPICS V flow cytometer/sorter.

B. Grazing experimentation:

We want to use many different phytoplankton cells simultaneously and monitor the rate of decrease of each upon exposure to a grazer(s). For the data here, we have selected 514 nm excitation using an argon-ion laser because it "favors" the phycoerythrin-containing cyanobacteria.

The important subsequent step is defining our axes. Here we desire maximum separation of the various chromatic groups of cells. After considerable experimentation we have settled on the use of DUN (*Dunaliella tertiolecta*, a chlorophyte) and DC-2 (*Synechococcus* sp., a phycoerythrin-containing cyanobacteria) grown under standard light, temperature and transfer conditions, as our biological standards/controls. This permits us to construct a defined map. Our map limits are thereby identified as to what is positioned where (Fig. 4). While other pigment and/or light scatter settings can be used for discrimination, this scheme relies on fluorescence only and has been designed for grazing and natural population studies, with five useful regions. Region A is for moderate chlorophyll per cell and no phycoerythrin or phycocyanin. Organisms which fall here include: *Dunaliella tertiolecta*, (DUN); *Phaeodactylum tricornutum*, (Phaeo); and *Isochrysis*, (Iso). Region E is for strong chlorophyll per cell and includes: *Heterocapsa triguertra* (HT 984), *Gonyaulax tamarensis*, (GT-PP); *Gonyaulax tamarensis* var. *excavata*, (GT 429) falls in region D, perhaps due to green fluorescing luciferin; Region B is where cells with major phycoerythrin fluorescence fall, namely the cyanobacterium DC-2. Region C is the cryptomonad niche, where *Chroomonas salinas* (3C) falls because it has both chlorophyll and phycoerythrin fluorescence.

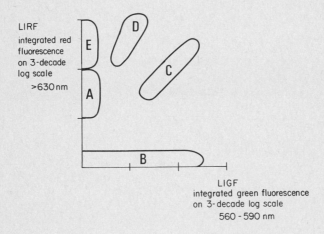

Figure 4. General map locations available using DUN (*Dunaliella tertiolecta*, a chlorophyte) as a biological standard for red fluorescence and DC-2 (*Synechococcus* sp., a cyanobacterium) as a biological standard for green fluorescence.

Numbers of cells at each map location can be integrated by the flow cytometer yielding relative numbers. To obtain actual concentrations per ml, we must add standard beads at known concentrations or do direct counts. If we increase the gains or laser power, the signal moves away from the origin. Once these settings are established, they cannot be altered if intercomparison of data is desired.

There are many possible alternative choices, such as red fluorescence vs. forward angle light scatter, etc. At the moment, forward angle light scatter is problematic in that it is measured by a photodiode and is limited to a linear vs. three-decade log scale. Installation of a log amplifier should remedy this current shortcoming. Biomedical researchers have, for the most part, worked with cells with very similar light scatter; they have not had to deal with the diversity of size and shape which challenges biological oceanographers.

Real data are presented in Figure 5. Mixtures of clonal phytoplankton cultures were analyzed using a Coulter EPICS V single beam flow cytometer based on the variation of autofluorescing pigments naturally contained in these cells. Individual filter feeding bivalve mollusks, the blue mussel, *Mytilus edulis*, were placed in two

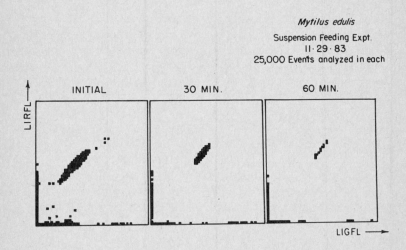

Figure 5. Initial, 30 min and 60 min culture mixture data prior to start of grazing experiments (Cucci et al., in prep.). Phytoplankton used were position A = Platy 1 (*Tetraselmis* sp., a prasinophyte); B = DC-2 (*Synechococcus* sp., a cyanobacterium); C = 3C (*Chroomonas salina*, a cryptomonad). Both LIGFL and LIRFL are displayed on a three decade log scale.

liter beakers containing these cell suspensions and the suspensions monitored with time. In addition, feces were collected after 2 h and gently sonified until dissociated and then analyzed. Data analysis of bivariate contour plots of the fluorescence signals permits assessment of filtration rates, ingestion rates, utilization and assimilation rates (Cucci et al., in prep.).

C. Automated characterization of natural populations:

Natural populations acquired by pumping water from subsurface particle maxima (maxima identified by transmissometry or *in vivo* chlorophyll fluorometry) can be analyzed either as a raw water sample or by concentrating larger particles between Nitex netting. For the data presented here, organisms were concentrated between 20-55 µm Nitex netting as pumped from a subsurface maximum at a station near Monhegan Island ($43°15'N$ $69°15'W$) into nets submerged in a water bath to reduce friction on the cells. Figure 6 is a bivariate three-dimensional plot of linear forward angle light scatter on the x-axis and log (three decades) of integrated red fluorescence (>630 nm emission ≅ chlorophyll). Twenty-five thousand events were analyzed. While relative numbers of cells in different subpopulations can be evaluated, these data are not useful without identification of the subpopulations into taxonomic, or better yet, functional groups. To

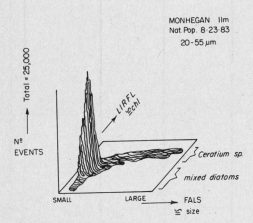

Figure 6. Flow cytometric analysis of FALS vs. Log Integrated Red Fluorescence of natural population from subsurface chlorophyll maximum (11 meters) near Monhegan Island, 22 km from Boothbay Harbor, coastal Gulf of Maine. The sample was concentrated between 22-50 µm plankton netting by pumping seawater through plankton nets assembled in a water bath to reduce friction on the cells.

accomplish this, cells from each subpopulation were sorted for identification. The low light scatter (FALS) and low fluorescent (LIRFL) subpopulations were identified as a mixture of diatoms, but the high fluorescent subpopulation was clearly unialgal *Ceratium* sp. ($\cong$98%). This suggests the ability to separate chromatic groups to establish functional relationships in nature.

As summarized by Sakshaug (1980) in his review chapter entitled Problems in the Methodology of Studying Phytoplankton: "In phytoplankton ecology, in particular when dealing with chemical methods, it is crucial to separate groups of organisms from each other and from detritus (dead organic matter) so that chemical data can be assigned correctly to the various groups of organisms and a correction made for detritus. This is one of the most difficult methodological problems in phytoplankton ecology... Problems increase with maturity of a community."

Flow cytometry and fluorescence-activated cell sorting may provide a means by which rapid analysis and separation of phytoplankton is feasible at sea and thus lead us to a better understanding of the organisms in the oceanic environments.

ACKNOWLEDGEMENTS

I thank O. Holm-Hansen for the opportunity to participate in this interesting symposium and helpful comments on an earlier version of this manuscript. Collectively, the authors thank P.K. Horan and K. Muirhead for the introduction to flow cytometers/sorters. Funding for the instrument at Bigelow Laboratory was through NSF OCE 82-13567 and ONR N00014-81-C-0043. Partial support was provided by OCE 81-21331 and NA-82-FA-C-00043.

P. Boisvert prepared the manuscript and J. Rollins and K. Knowlton prepared the illustrations. This is Bigelow Laboratory contribution number 84001.

REFERENCES

Cucci T.L., Shumway S., Selvin R., Guillard R.R.L. and Yentsch C.M. (in prep.) Approaching suspension-feeding experimentation via flow cytometry.

Horan P.K. and Wheeless Jr. L.L. (1977) Quantitative single cell analysis and sorting. Rapid analysis and sorting of cells is emerging as an important new technology in research and medicine. Science 198 : 149-157.

Kruth H.S. (1982) Flow cytometry: Rapid biochemical analysis of single cells. Anal Biochem 125 : 225-242.

Melamed M.R., Mullaney P.F. and Mendelson M.L. (1979) Flow cytometry and sorting. John Wiley and Sons, NY. 716 pp.

Olson R.J., Frankel S.L., Chisholm S.W. and Shapiro H.M. (1983) An inexpensive flow cytometer for the analysis of fluorescence signals in phytoplankton:chlorophyll and DNA determinations. J exp mar Biol Ecol 68 : 129-144.

Sakshaug E. (1980) Problems in the methodology of studying phytoplankton. In : Morris, I. (ed.) The Physiological Ecology of Phytoplankton, Blackwell, London, 57-91.

Trask B.J., van den Engh G.J. and Elgershuizen J.H.B. (1982) Analysis of phytoplankton by flow cytometry. Cytometry 2(4) : 258-264.

Yentsch C.M., Horan P.K., Muirhead K., Dortch Q., Haugen E., Legendre L., Murphy L.S., Perry M.J., Phinney D.A., Pomponi S.A., Spinrad R.W., Wood M., Yentsch C.S. and Zahuranec B.J. (1983a) Flow cytometry and cell sorting: a powerful technique for analysis and sorting of aquatic particles. Limnol Oceanogr 28(6) : 1275-1280.

Yentsch C.M., Mague F., Horan P.K. and Muirhead K. (1983b) Flow cytometric analysis of DNA in the dinoflagellate *Gonyaulax tamarensis* var. *excavata*. J exp mar Biol Ecol 67 : 175-183.

Yentsch C.M., Cucci T.L. and Phinney D.A. (submitted) Photoadaptation of *Gonyaulax tamarensis* to low photon flux densities: chloroplast photomorphogenesis.

Yentsch, C.S. and Yentsch C.M. (1979) Fluorescence spectral signatures: the characterization of phytoplankton populations by the use of excitation and emission spectra. J Mar Res 37 : 471-483.

Yentsch C.M. and Yentsch C.S. (in press) Emergence of optical instrumentation for measuring biological properties. In : Barnes M. (ed.) Oceanography and marine biology: an annual review.

DETERMINATION OF ABSORPTION AND FLUORESCENCE
EXCITATION SPECTRA FOR PHYTOPLANKTON

B.G. MITCHELL and D.A. KIEFER

Department of Biological Sciences, University of Southern
California, Los Angeles, California 90089, USA.

I. INTRODUCTION

Investigations of photosynthetic processes in both aquatic ecosystems and in the laboratory have sought to couple measurements of the rates of production with quantitative assessment of the ambient light field and the absorption capacity of the cells. Kiefer etal. (1979) and Bricaud etal. (1983) have described techniques for determination of absorption coefficients for cultures. For field studies, however, where the cell concentration is very dilute, direct measurement of absorption is not possible. Most field studies have assumed a constant absorption per chlorophyll, and have estimated the amount of absorbed light from measurements of the chlorophyll concentration and the available light for photosynthesis (Rhode, 1965; Tyler, 1975; Dubinsky and Berman, 1976; Morel, 1978). We present here a technique for measuring the absorption coefficient (a_p) for cultures or field samples. Yentsch (1957) first applied the technique of direct measurement of absorption on filters for algal cultures, and for qualitative analysis of suspended marine particulates (1962). Faust and Norris (1982) have used derivative absorption spectroscopy of cultured phytoplankton on glass fiber filters in order to assess pigment concentration. The problem with measurement on filters is that the optical environment has been modified and corrections for the effects this has on the measurement must be made if quantitative applications are desired. Kiefer and SooHoo (1982) have described a correction which estimates the amplification effect due to scattering by glass fiber filters. A more accurate technique for correcting this effect is described here.

In addition to making routine measurements of a_p in the field and lab, we have also developed an inexpensive and simple spectrofluorometer capable of rapidly measuring fluorescence excitation spectra of chlorophyll <u>a</u> contained in particles retained on filters. Yentsch and Yentsch (1979) used spectra which were uncorrected for the system response to describe the general taxonomic composition of the phytoplankton in different oceanic regions. By correcting for the system response, and reabsorption of fluoresced light we have extended this technique to studies of primary production and photoadaptation.

II. MATERIAL AND METHODS

A. Sample preparation

Samples from both the field and lab were filtered onto Whatman GF/C filters with low vacuum pressure (<5mm.Hg.). Field samples were collected with Niskin bottles on a rosette containing a CTD package. For laboratory analyses described here, cultures of Dunaliella tertiolecta were grown in f/2 enriched sterile seawater under continuous illumination from cool white fluorescent lamps. Samples were kept dark and refrigerated before analysis of fluorescence spectra, which was done as soon as possible after filtration. Prior to absorption spectra analysis, the filters were moistened with filtered seawater to ensure saturation, since the optical density of the filter varied with the degree of saturation. Storage of the samples at -20 C. prior to measurement of absorption can be done with no apparent effect, however samples should not be frozen prior to fluorescence analysis because a loss of energy transfer from accessory pigments to chlorophyll a is apparent. Nevertheless, the losses due to freezing are comparable for different samples so that qualitative comparisons of spectral shapes can still be used to obtain information about photoadaptation (Neori etal. 1982).

B. Instrumentation

For measurements of particle absorption, a vertical light path sample compartment (Butler,1962), specifically described by Kiefer and SooHoo (1982) is used. The components include a Bausch and Lomb high intensity monochromator(33-86-76) coupled to a Bausch and Lomb 45 watt tungsten-halide light source. The optical density for the sample, $OD_S(\lambda)$, is calculated from measurements of the light intensities transmitted by a sample filter, $E_S(\lambda)$, and a blank filter $E_B(\lambda)$. Measurements are made with a Gamma 20-20-10 photomultiplier and 2900 autophotometer. The photometer is interfaced, after amplification, to a Digital Equipment Corporation PDP 1103 with which the analog data is digitized, averaged, smoothed, and processed.

For fluorescence excitation spectra, the mirror which reflects the monochromatic beam onto the sample is replaced with a long wavelength transmitting dichroic filter(OCLI CSF-A) that reflects the excitation beam onto the sample while allowing the chlorophyll a fluorescence to

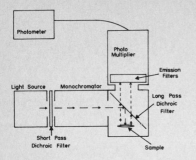

Figure 1.
A schematic diagram of the optical system configured for fluorescence excitation spectra.

pass to the photomultiplier which is mounted above the dichoric mirror. A 680 nm. bandpass filter (Melles-Griot 03-fiv-065) is placed before the photomultiplier to discriminate between fluoresced and reflected light (Fig. 1). This optical configuration facilitates system calibration and thereby allows the generation of system corrected fluorescence excitation spectra. A system spectrum is obtained by measuring the fluorescence of the dye rhodamine-B, dissolved in ethylene glycol at a concentration of 3 g./l. At such high concentrations, this dye is a quantum counter which absorbs all photons and emits them with a constant quantum efficiency in a broad band above 600 nm. (Melhuish,1962; Taylor and Demas,1979). The system spectrum is used in the fluorescence correction algorithm. Spectral limitations of the dichroic filter restrict measurements to excitation wavelengths less than 560 nm.

C. Correction and analysis of spectra

To minimize seawater requirements, in the field we routinely make measurements of absorption with two layers of folded or stacked sample filters. This requires a two step correction in order to derive a_p: a correction from 2 to 1 layers and from 1 layer to the absorption coefficient in a non-scattering medium. Correction algorithms for both steps were determined empirically using cultures of D. tertiolecta.

For the first step, the optical densities for a single and double layer were compared. The second step requires measurement of the absorption of a single glass fiber filter and the true absorption in a minimally scattering medium. In order to measure the latter value, the refractive index of an exponentially growing culture was matched with a solution of bovine serum albumin (BSA) (Barer,1955). By minimizing the scattering and maximizing the angle of acceptance of the photomultiplier, measurement of a value very close to the volume absorption coefficient is possible. To determine the BSA concentration which best matched the refractive index of the cells, D. tertiolecta cells in exponential growth were collected by centrifugation and resuspended in a series of dilutions of a stock solution of 50% BSA(w/v) in distilled water. The dilution which transmitted maximally at 550 nm. was used to measure a_p. A diffusing plastic disk was placed at the photomultiplier surface according to the Shibata (1958) technique.

For convenience, fluorescence excitation spectra are determined for the same samples as absorption. Since a relatively large volume of seawater must be filtered for a determination of a_p, measurements of fluorescence on the same sample are subject to serious quenching by reabsorption and attenuation of the light field to which the particles are exposed. The correction algorithm for these two phenomena were determined by filtering from 1 to 50 mls. of an exponentially growing culture of D. tertiolecta onto GF/C filters. Fluorescence and absorption spectra were determined for each sample. Also, determination of the optical density of a blank single layer GF/C filter, relative to air, was carefully made by placing the filter directly on the diffusing plastic resulting in a distance of only about 3 mm. from the surface of the filter to the photocathode film on the photomultiplier entrance window. This ensured that all forward scattered light was accepted by the photomultiplier.

III. RESULTS

A. Volume absorption coefficient algorithm

For our routine measurements of absorption for a double layer of a sample filter which has been cut in half and stacked, the first correction which must be applied to the raw data is to generate a true spectrum of the equivalent material on a single layer. According to Beers Law for absorbing materials in pure solution, one expects a doubling of the optical density when the geometric path length is doubled. However, since the glass fiber filters used in this analysis are highly diffusing, the concepts for pure solutions are not applicable. In fact, because of the large scattering cross-section of the filters, the actual optical pathlength of the sample is greater than the geometric pathlength. Butler (1962) defined the term β as the ratio of the optical to geometric pathlength. Stavn (1981) defined this parameter as the mean pathlength and used it to describe absorption of light in scattering media. For our measurements on glass fiber filters, we have divided the correction into two parts: the correction for $\beta(\lambda,1)$ for a single layer relative to a minimally scattering suspension, and $\beta^*(\lambda,2)$ for a second layer relative to a single layer. $\beta^*(\lambda,2)$ is not a true β in the sense of Butler(1962) because both the single and double layer measurements are concerned with diffuse transmittance, while Butler was concerned with the amplification for a diffuse transmittance relative to a collimated transmittance.

For E_S and E_B, the measured transmitted irradiance for the sample and blank, respectively, the optical density for the sample is defined as:

$$OD_S(\lambda,n) = -\log_{10}[E_S(\lambda,n)/E_B(\lambda,n)] \qquad \text{Eq. 1}$$

where λ is the wavelength, and n is the numbers of filter stacks. Defining X to be the ratio of the volume of sample filtered to the clearance area of the filter, then in general the volume absorption coefficient for particulates can be determined by equation 2:

$$a_p(\lambda,n) = \frac{2.3 \cdot OD_S(\lambda,n)}{X \cdot \beta(\lambda,n)} \qquad \text{Eq. 2}$$

The constant 2.3 is necessary to convert $\log_{10}$ to natural log units. For the measurements we routinely make with two layers,

$$\beta(\lambda,2) = \beta(\lambda,1) \cdot \beta^*(\lambda,2) \qquad \text{Eq. 3}$$

Operationally, $\beta^*(\lambda,2)$ is defined by the OD for two layers divided by two times the OD for one layer:

$$\beta^*(\lambda,2) = \frac{OD_S(\lambda,2)}{2 \cdot OD_S(\lambda,1)} \qquad \text{Eq. 4}$$

When the nature of this relationship was studied with exponentially growing cells, a clear dependence on the sample OD was found. The results of this analysis for Whatman GF/C filters are illustrated in figure 2, which is a composite of analyses done for two different volumes of culture filtered so that a broad range of OD could be achieved. In principle β^* cannot be less than 1 and the solid line in figure 2 defined by equation 5 provides an adequate description of the data.

$$\beta^*(\lambda,2) = 1.0 + 0.325 \cdot [OD_S(\lambda,2)^{-0.561}] \qquad \text{Eq. 5}$$

Duntley(1942) investigated the general principles of optical properties of diffusing materials, and derived equations which describe the relationships for the empirical results which we have observed. The dashed line in figure 2 represents the results of the predictions of the Duntley equations, using as input the absolute OD of a blank single and double GF/C filter stack, relative to air.

$\beta(\lambda,1)$ in equation 3 is operationally defined as the ratio of the absorption of cells measured on a single filter layer to the same equivalent pathlength of cells suspended in a solution of bovine serum albumin (BSA). The results of the analysis to determine this parameter for GF/C filters are illustrated in figure 3. The right hand panel of figure 3 demonstrates that $\beta(\lambda,1)$ is also a function of the sample OD. The data in this experiment are, however, adequately described by a linear equation. In terms of the OD for a single layer:

$$\beta(\lambda,1) = 2.26 - 1.5 \cdot OD_S(1) \qquad \text{Eq. 6}$$

B. **Chlorophyll a fluorescence excitation algorithm**

The uncorrected fluorescence excitation spectrum of phytoplankton will be proportional to the absorption bands which are effective in transferring energy to the fluorescing molecules, the efficiency of transfer of the absorbed energy, the fluorescence efficiency of the fluorescing molecules, and the intensity of the excitation beam. The significance of the system excitation energy on the spectrum can be appreciated by examining figure 4 which shows an uncorrected sample spectrum, the system spectrum determined with rhodamine-b, and the sample spectrum corrected for spectral properties of the system alone. The corrected spectrum is dramatically different from the uncorrected spectrum, and it is especially important to note that the chlorophyll a bands at 380 nm. and 435 nm. become more prominent in the corrected spectrum.

For an optically thin sample, the correction depicted in figure 4 is sufficient. In general the uncorrected spectrum is described by equation 7:

$$Fl_u(\lambda) \propto \phi_{fl}(\lambda) \cdot A(\lambda) \qquad \text{Eq. 7}$$

where $Fl_u(\lambda)$ is the uncorrected fluorescence signal, $\phi_{fl}(\lambda)$ is the fluorescence efficiency for absorbed light, and $A(\lambda)$ is the absorbed light. The proportionality in this equation and equation 12 below is appropriate because only a fraction of the fluoresced light is detected. For an optically thin sample where the emergent beam is approximately equal to the incident beam $A(\lambda)$ can be described by equation 8:

$$A(\lambda) = E(\lambda) \cdot OD_S(\lambda) \qquad \text{Eq. 8}$$

where $OD_S(\lambda)$ is the sample optical density and $E(\lambda)$ is the incident light intensity. However, because the samples are not optically thin, we have replaced the incident intensity with the mean light intensity: $\bar{E}$. This value is a function of the incident intensity, the OD_S for the sample, relative to a blank, plus the OD_B for a blank filter relative to air. Integrating Beers Law:

$$\bar{E}(\lambda) = E(\lambda)\frac{[1-10^{-(OD_B + OD_S)}]}{(OD_B + OD_S)} \qquad \text{Eq. 9}$$

The spectral distortions which are caused by a lack of correction for the mean light level can be seen in figure 5 which is a comparison of fluorescence excitation spectra for different volumes of the same

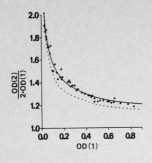
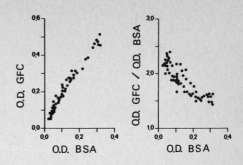

Figure 2.

$\beta*(\lambda,2)$ plotted as a function of the optical density of a single GF/C layer.

Figure 3.

The optical density of a sample of D. tertiolecta on a GF/C filter, and the corresponding $\beta(\lambda.1)$ plotted as functions of the optical density for the cells in Bovine Serum Albumin.

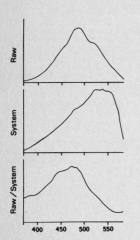
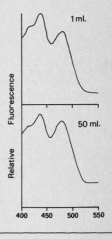

Figure 4.

Relative spectral fluorescence of a raw field sample, the system spectrum and the raw spectrum corrected for the system response.

Figure 5.

Relative spectral fluorescence for 1 ml. and 50 ml. of a culture of D. tertiolecta on a GF/C filter.

culture of Dunalliela which have been corrected only for the system spectral excitation energy. One can clearly see that in regions of the spectrum where absorption is low, there is an increase in the apparent fluorescence relative to regions where absorption is high. Specifically, at 480 nm. where there is less absorption than 435 nm., the peak in the 50 ml. sample is higher relative to the 435 nm. peak; also the low fluorescence region around 550 nm. is higher relative to both the 480 mm. and the 435 nm. peaks for the 50 ml. sample.

An additional consideration which must be made in order to calculate fluorescence spectra is the quenching of the fluoresced light by the chlorophyll a absorption band in the red. The quenching term, Q, is dependent on the fraction of the diffuse fluoresced light which is absorbed before leaving the filter. In the analysis for absorption we have determined the optical density of the sample for collimated light, and with appropriate scaling to account for the higher absorption for diffuse light relative to collimated light we use the OD_S (683,1) to estimate Q:

$$Q = 10^{-[OD_S(683,1) \cdot C]} \qquad \text{Eq. 10}$$

where C is the scaling factor which we have found to be approximately 2. One can now define a corrected fluorescence spectrum $Fl_C(\lambda)$ in terms of the above algorithm:

$$Fl_C(\lambda) = \frac{Fl_u(\lambda)}{\bar{E}(\lambda) \cdot Q} \qquad \text{Eq. 11}$$

so that:

$$\emptyset_{fl}(\lambda) \propto \frac{Fl_C(\lambda)}{OD_S(\lambda)} \qquad \text{Eq. 12}$$

Figure 6 illustrates, in the top panel, the overall consequences one can expect if correction for only the system excitation is considered. For increasing densities of <u>Dunaliella</u> on the filter, an initially linear region is followed by a region where quenching becomes dominant. When the algorithm described above is applied, the data become linearized, as seen in the lower panel of figure 6. The range within which we routinely work corresponds to the region above 20 ml., where $OD_S(\lambda,1)$ is greater than about 0.15 in the maximally absorbing regions of the Soret bands. Clearly, if one wants to compare spectra in more than a qualitative way, these corrections should be applied.

C. <u>Applications in field studies</u>

Figures 7, 8 and 9 show the results of application of the foregoing algorithms to a vertical profile taken at a station approximately 200 km. west of the southern Baja California peninsula. This station can be considered typical of stations in the transition zone between the California current and the oligotrophic central Pacific gyre. The mixed layer extended to a depth of 30 m., contained little chlorophyll, and a subsurface chlorophyll maximum was found at 70 m.

Figure 7 contains chlorophyll specific volume absorption coefficients which have units of $m^2/(mg.Chl\ \underline{a})$. Two features of spectral absorption are apparent. First, the magnitude for absorption in the blue region of the spectra is high in surface and deep water. The increase in the magnitude of the deep sample can be attributed to photoadaptation since there is an associated dramatic increase in the absorption in the

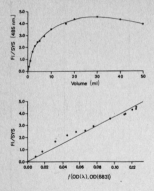

Figure 6

System corrected fluorescence at 485 nm. for D tertiolecta plotted as a function of volume filtered, and as a function of the mean light level and quenching algorithm.

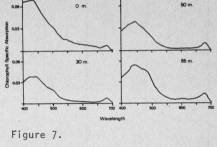

Figure 7.

Depth profile of the chlorophyll specific volume absorption coefficient, plotted against wavelength.

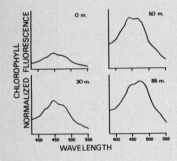

Figure 8.

Depth profile of the chlorophyll normalized corrected fluorescence, plotted against excitation wavelength.

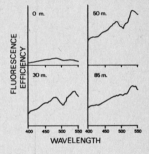

Figure 9.

Depth profile of the fluorescence efficiency, plotted against excitation wavelength.

---

accessory pigment bands between 470 nm. and 520 nm. Such spectral adaptation is seen to a lesser extent at 30 and 50 m. The high absorption for surface samples cannot be attributed to photoadaptation, since this population would be experiencing a much higher mean irradiance than the deeper samples. It is possible that cells in the mixed layer add protective carotenoid pigmentation (Krinsky,1968) to minimize the detrimental effects of high light levels including UV radiation (Smith etal., 1980). Second, the absorption in the blue region of the spectrum for field samples relative to the red peak (eg. 435:675 nm) is higher in field samples than is seen for healthy laboratory cultures. This seems to be generally true, though more striking in surface and very deep samples, and was reported originally by Yentsch (1962).

The fluorescence excitation spectra for this station, normalized to chlorophyll a are presented in figure 8. Spectral shifts attributable to the addition of accessory pigments are again clearly present. Also, the magnitude of the fluorescence per chlorophyll increases with depth. Spectral shifts and increased magnitudes are also seen in the fluorescence efficiency spectra in figure 9. Additionally, the efficiency spectra clearly indicate that in all cases, light absorbed by accessory pigments is more efficient in exciting chlorophyll a fluorescence than light absorbed by chlorophyll a itself.

IV. DISCUSSION

We propose that the changes in magnitude and shape of the chlorophyll a excitation spectra are caused by photoadaptation of the viable phytoplankton. The fluorescence we measure at 683 nm. could be from either phaeopigments or chlorophyll a. However, the relatively small contribution at 416 nm. excitation, where phaeopigments absorb maximally, implies that phaeopigment fluorescence is unimportant in interpreting these spectra. Even in the bottom of the euphotic zone where the fluorometrically determined chlorophyll:phaeopigment ratio is about one, this appears to be a valid assumption. If the degradative processes which convert chlorophyll to phaeopigment are sufficiently harsh, they may disrupt the energy transfer coupling between the accessory pigments and any residual fluorescing molecules.

Yentsch and Yentsch (1979) have proposed that changes in uncorrected excitation spectra are attributable to taxonomic changes. It is certainly true that for unialgal cultures, the shapes of the spectra can distinguish differences in taxonomic groups. However, when one induces photoadaptation in cultures, the results are generally similar: both accessory pigments absorption, and their contribution to chlorophyll a fluorescence increase. The accessory pigments absorb predominantly between 460 nm. and 550 nm. although at wavelengths longer than 500 nm., the fucoxanthin in diatoms and peridinin of dinoflagellates are the main accessory pigments. In open ocean stations as the one presented here, these taxa are relatively rare so that the accessory pigments absorption bands for the taxa present are mostly in a narrow range from 460-490 nm. Thus, taxonomic discrimination would be difficult using these absorption bands, and any photoadaptive effects would act cumulatively to produce similar spectral shifts. Furthermore, changes comparable to those presented here have been induced in field cultures incubated under light limiting conditions (Neori etal.,1982; Mitchell and Kiefer,1983). Changes were seen within one day, a time period too short for significant taxonomic changes to occur. Also, the light was attenuated with neutral density filters, indicating that light intensity, and not spectral composition is the relevant parameter. We have seen similar patterns in nutrient rich Antarctic waters as well as the oligotrophic Pacific gyre where nutrients were not measurable in the euphotic zone. Thus, nutrient availability is not a cause of the spectral shifts.

Although it is probably valid to argue that the fluorescence spectra are due to photosynthetically viable pigments, the absorption spectra include detrital pigments as well as any non-photosynthetic pigments which may be produced by the phytoplankton. Interpretation of these spectra is therefore more complicated. In figure 7, the samples below the mixed layer clearly show changes which one could argue are photoadaptive. However, the mixed layer shows the highest overall chlorophyll specific volume absorption coefficient, while the spectrum is dominated by short wavelength absorption, compared to the deeper samples. This high absorption clearly does not transfer energy efficiently to chlorophyll $\underline{a}$, as evidenced by the fluorescence excitation and efficiency spectra (Figs. 8,9). It is reasonable to propose that these features are associated with photo-protective pigments, since the mixed layer has high irradiance levels, and photoinhibition is known to occur at high light levels (Jassby and Platt,1976). An alternative hypothesis is that there is a higher accumulation of detrital particulates in the mixed layer. Yentsch (1962) proposed that the high value in the ratio of blue to the red absorption peak in field samples, as compared to cultures, might be due to the presence of detrital particles. He noted for very deep samples the absorption is dominated by short wavelength absorption and the red peak disappears. We have found for absorption spectra of mixed layer samples from the central Pacific gyre, the ratio of blue absorption to the red peak is greater than the example presented here. Furthermore, the water column in the gyre is characterized by having the highest primary production in the mixed layer, long term stability in vertical profiles of particle and chlorophyll concentrations, and nitrogenous nutrients are undetectable in the euphotic zone. The observed photosynthesis in the mixed layer must be balanced by grazing for long term stability to exist, and is in turn maintained by recycled nutrients (Eppley etal., 1973). Such a mechanism could result in a higher proportion of detrital particles in the mixed layer compared to the lower euphotic zone, provided grazing is carried out predominantly by microzooplankton producing non-sinking fecal material.

Application of these techniques for quantitative analysis of light absorption together with measurements of photosynthesis has allowed assessment of the dependence of photosynthetic quantum efficiency on the irradiance level of growth (Mitchell and Kiefer,1983). We found that photoadaptation can lead to a three fold increase in quantum efficiency for exponentially growing, nutrient saturated field cultures incubated at different light levels. This is consistent with our model of photosynthesis (Kiefer and Mitchell,1983). We are currently analyzing *in situ* quantum efficiency using these techniques and preliminary results indicate that deep euphotic zone samples have higher efficiencies. Obviously, absorption by non-photosynthetic pigments in the natural samples will cause interpretive problems; nevertheless, knowledge of total ecological efficiency of photosynthesis can be determined and could be very useful for satellite image analysis.

These quantitative corrections for absorption and fluorescence spectra offer a new tool with which to investigate phytoplankton photosynthetic processes in both the laboratory and field. The absorption coefficient technique can be accomplished using a spectrophotometer interfaced with a microcomputer, equipment commonly available even in routine field experiments. With efficient computer programming, spectral absorption coefficients can be determined more easily than a fluorometric analysis of extracted chlorophyll.

V. REFERENCES

Barer R (1955) Spectrophotometry of clarified cell suspensions. Sci. 121:709-715.

Bricaud A, A Morel and L Prieur (1983) Optical efficiency factors of some phytoplankters. Limnol.Oceanogr. 28:816-832.

Butler W (1962) Absorption of light by turbid materials. J.Opt.Soc. Am. 52:292-299.

Dubinsky Z and T Berman (1976) Light utilization efficiencies of phytoplankton in Lake Kinneret(Sea of Galilee). Limnol.Oceanogr. 21: 226-230.

Duntley SQ (1942) Optical properties of diffusing materials. J.Opt.Soc. Am. 32:61-69.

Faust MA and CH Norris (1982) Rapid *in vivo* analysis of chlorophyll pigments in intact phytoplankton cultures. Br.Phycol.J. 17:351-361.

Eppley, RW, EH Ringer, EL Venrick and MM Mullin (1973) A study of plankton dynamics and nutrient cycling in the central gyre of the North Central Pacific Ocean. Limnol.Oceanogr. 18:534-551.

Jassby AD and T Platt (1976) Mathematical formulation of the relationship between photosynthesis and light for phytoplankton. Limnol.Oceanogr. 21:540-547.

Kiefer DA, RJ Olson and WH Wilson (1979) Reflectance spectroscopy of marine phytoplankton. Part I. Optical properties as related to age and growth rate. Limnol.Oceanogr. 24:664-672.

—and JB SooHoo (1982) Spectral absorption by marine particles of coastal waters of Baja California. Limnol.Oceanogr. 27:492-499.

—and BG Mitchell (1983) A simple, steady-state description of phytoplankton growth based on absorption cross section and quantum efficiency. Limnol.Oceanogr. 28:770-776.

Krinsky NI (1968) The protective function of carotenoid pigments. In: AC Giese (ed.) Photophysiology, vol. III. Academic Press, New York. pp. 123-195.

Melhuish WH (1962) Calibration of spectrofluorometers for measuring corrected emission spectra. J.Opt.Soc.Am. 52:1256-1258.

Mitchell BG and DA Kiefer (1983) Phytoplankton photosynthetic quantum efficiency and spectral fluorescence excitation efficiency response to light limitation. Eos 63:961.

Morel A (1978) Available, usable and stored radiant energy in relation to marine photosynthesis. Deep-Sea Res. 25:673-688.

Neori A, BG Mitchell, J SooHoo, DA Kiefer and O Holm-Hansen (1982) Increased efficiency of energy transfer from accessory pigments to chlorophyll a in phytoplankton with depth: An adaptation to changing light conditions or reaction to environmental stress. Eos 63:96.

Rodhe W (1965) Standard correlations between pelagic photosynthesis and light. In: CR Goldman (ed.) Primary productivity in aquatic environments. Mem.Ist.Ital.Idrobiol. 18(Suppl):365-381.

Shibata K (1958) Spectrophotometry of intact biological materials. Absolute and relative measurements of their transmission, reflection and absorption spectra. J.Biochem. 45:599-623.

Smith RC, KS Baker, O Holm-Hansen and R Olson (1980) Photoinhibition of photosynthesis in natural waters. Photochem.Photobiol. 31:585-592.

Stavn RH (1981) Light attenuation in natural waters: Gershun's law, Lambert-Beer law, and the mean light path. Appl.Optics 20:2326-2327.

Taylor DJ and JN Demas (1979) Light intensity measurements I: Large area bolometers with microwatt sensitivities and absolute calibration of the rhodamine-B quantum counter. Anal.Chem. 51:712-717.

Tyler JE (1975) The in situ quantum efficiency of natural phytoplankton populations. Limnol.Oceanogr. 20:976-980.

Yentsch CS (1957) A non-extractive method for the quantitative estimation of chlorophyll in algal cultures. Nature 179:1302-1304.

—(1962) Measurement of visible light absorption by particulate matter in the ocean. Limnol.Oceanogr. 7:207-217.

—and CM Yentsch (1979) Fluorescence spectral signatures: The characterization of phytoplankton populations by use of excitation and emission spectra. J.Mar.Res. 37:471-483.

Acknowledgements

This work was supported by grants 04-7-158-44123 from the National Oceanic and Atmospheric Administration and n°0014-81-K-0388 from the Office of Naval Research.

SUBJECT INDEX

Acropora granulosa	73
Arsenophospholipids	55
Algal exudates	49
Amino acids	63
Ammonium transport rate	116
Antarctic ocean	28
Antarctic phytoplankton	19
Anthosphaera quadricornu	92
Arsenic detoxification	55
Arsenobetaine	56
Assimilation number	25
Asterionella glacialis	92
Bacterial secondary production	45
Bacterioplankton	45
Barents sea	2
Beaufort sea	7
Bering sea	2
Bering strait	3
Cacodylate	57
Calcidiscus leptoporus	92
Cage culture	113
Carbohydrates	63
Carbon-limited growth	39
Cell elutriation	145
Cell sorting	141
Chaetoceros sp.	62
Chaetoceros affinis	63
breve	92
compressus	119

curvisetrum	92
decipiens	92
gracilis	57
simplex	94
tortissimus	10
Chorella pyrenoidosa	38
Chlorophyll	107, 143
a	21, 92, 118, 129
b, c	129
fluorescence	157
Chromatic adaptation	137
Chroomonas salinas	151
Coccolithophores	130
Coccolithrophorids	92
Cryptomonads	131
Cyanobacteria	130
3', 5'-cyclic AMP (cAMP)	47
Diel P/R ratios	73
Denmark strait	6
Density gradient centrifugation	145
Detonula confervacea	14
Detritus	144
Dialysis chambers	113
Diatoms	92, 113, 130
Dimethylarsenosoribosides	55
Dinoflagellates	92, 95, 130
Dissolved organic matter	45, 61
Downwelling attenuation coefficient	105
Dunaliella tertiobecta	151, 157
Electrons flow	36
Emiliania huxleyi	92
Energy conversion	
- efficiency of	40
Energy transfer	148
Epifluorescent microscopy	135
Epontic algae	7
Escherichia coli	51
Euphausia superba	29

Euphotic zone	21
Excretion	61
Flow cytometry	141
Fluorescence techniques	143
chlorophyll	157
Fluorescent signatures	129
staining	148
Fucoxanthin	130
Glucose uptake	46
Glycollate	63
Gonyavlax tamarensis	150
Great Barrier Reef	73
Greenland sea	2, 6
Growth rate	65, 113
Gulf of Mexico	118
Gulf of Naples	89
Gulf of Salerno	89
Gymnodinium microadriaticum	73
Heterocapsa triquertra	151
Ice edge	8
"island-mass" effect	27
Isochrysis sp.	151
Leptocylindrus danicus	94
minimus	119
Light	4, 8, 23
Light -limited growth	35
saturation curves	73
shade adaptation	9
Mediterranean sea	101
Microzones	
of high concentrations	46
Microzooplankton	144
Multiphasic transport system	46
Mytilus edulis	152
	30

Nanoplankton	30
Nitrogen transport rate	115
<u>Nitzchia pungens</u>	119
<u>closterium</u>	92, 119
Norwegian Coast	2, 7
N/P ratio	65
Nutrient depletion	8
levels	3, 62
limitation	13
limited growth	35
salts	26
Ocean color scanner	138
Organophosphorus	49
Oxygen	92
<u>Oxytoxum variabile</u>	92
Peptides	63
Peridinin	130
<u>Peridinium quinquicorne</u>	119
Periplasmic space	51
<u>Phaeocystis pouchetti</u>	21
<u>Phaeodactylum tricornutum</u>	151
Phosphate transport rate	116
Photo-inhibition	24
Photosynthesis	73, 157
Photosynthetic electron transport	36
pigments	129
Phycobilin	131
Phycocyanin	130
Phycoerythrin	130
Phytoplankton	94, 114
Phytoplankton growth thermodynamics	35
Planktonic algae	61
Polar Front	2, 13, 28
Polysaccharides	48
extracellular	65
Primary production	6, 20, 45, 94, 109
<u>Prorocentrum balticum</u>	92
<u>minimum</u>	117, 119

Proteins	48
P vs I curves	11
Quantum efficiency	38, 157
Remote sensing techniques	138
Repellents	49
Ribulose 1,5-diphosphate carboxylase	39
Rhizosolenia alata	92, 122
Salmonella typhimurium	51
Scotia sea	5
Secondary production (bacterial)	45
Skeletonema costatum	62, 94, 119
Southern ocean	19
Specific growth rates	26
Stability erosion	8
Synechococcus sp.	151
Syracosphaera pulchra	92
Tampa Bay	118
Temperature	26
Thalassiosira decipiens	94
gravida	64
nordenskioeldii	14
tumida	20
weissflogii	41
Thalassiothrix fravenfeldii	92
Tidal mixing	133
Trace elements	27
Trondheimsfjord	4
Trimethylarsonium derivative	55
5-trimethylarsonium-ribosylglycerol-sulfate	57
Umbellosphaera tenuis	92
Water column stability	28
Weddell sea	3
Wind drift	3

Lecture Notes on Coastal and Estuarine Studies

Managing Editors:
R.T.Barber, M.T.Bowman,
C.N.K.Mooers, B.Zeitzschel

Springer-Verlag
Berlin
Heidelberg
New York
Tokyo

Volume 1
Mathematical Modelling of Estuarine Physics
Proceedings of an International Symposium Held at the German Hydrographic Institute, Hamburg, August 24–26, 1978
Editors: **J.Sündermann, K.-P.Holz**
1980. 119 figures, 1 table. VIII, 265 pages
ISBN 3-540-09750-3

Contents: Basic formulations and algorithms. – Tides and storm surges. – Baroclinic motions and transport processes.

Volume 2
D P. Finn
Managing the Ocean Resources of the United States: The Role of the Federal Marine Sanctuaries Program
1982. IX, 193 pages. ISBN 3-540-11583-8

Contents: Introduction. – Case Studies. – Interagency Coordination for the Management of Marine Resources. – The Marine Sanctuaries Program. – The Role of Designating Marine Areas for Special Management. – Recommendations and Conclusions. – Notes. – Alphabetical List of Major References.

Volume 3
Synthesis and Modelling of Intermittent Estuaries
A Case Study from Planning to Evaluation
Editors: **W.R.Cuff, M.Tomczak, Jr.**
1983. VIII, 302 pages. ISBN 3-540-12681-3

The Port Hacking Estuary Program set out in 1973 is analyzed as a case study of model-guided, multidisciplinary ecosystem studies. Organizational difficulties are discussed with the aim of helping future studies of this genre avoid similar mistakes. The book illustrates how a data set that is incomplete from the viewpoint of constructing a predictive dynamic ecosystem model can still be synthesized and used.

Lecture Notes on Coastal and Estuarine Studies

Managing Editors:
R. T. Barber, C. N. K. Mooers, M. J. Bowman, B. Zeitzschel

Springer-Verlag
Berlin
Heidelberg
New York
Tokyo

Volume 4
H. R. Gordon, A. Y. Morel
Remote Assessment of Ocean Color for Interpretation of Satellite Visible Imagery
A Review
1983. V, 144 pages. ISBN 3-540-90923-0

Contents: Introduction. – The Physics of Ocean Color Remote Sensing: Irradiance Ratio and Upwelling Subsurface Radiance, Atmospheric Effects. – In-Water Algorithms: The Phytoplankton Pigment Algorithms. The 'K' Algorithms. The Seston Algorithms. The Analytic Algorithm. Relationship Between the Algoirithms. – Atmospheric Correction. – Application of the Algorithms to CZCS Imagery. – Summary and Conclusions. – Appendix I: The Coastal Zone Color Scanner (CZCS). – Appendix II: Recent Developments: Clear Water Radiance Concept. Accuracy of Pigment Estimates. Applications. – References.

Volume 5
D. C. L. Lam, C. R. Murthy, R. B. Simpson
Effluent Transport and Diffusion Models für the Coastal Zone
1984. IX, 168 pages. ISBN 3-540-90928-1

Contents: Introduction. – Parameterization of Advection and Diffusion Processes. – Mathematical Models with Analytical Solutions. – Marching Technique Solutions for Straight Plume Equations: Effects of Scale Dependent Diffusivity. – Fully Two-Dimensional Computational Technique for Steady Plume Modelling. – Testing Finite Element Plume Using Examples with Analytical Solutions. – Vertification and Application. – References. – Appendix I: Notations. – Appendix II: Subject Index.

Volume 6
Ecology of Barnegat Bay, New Jersey
Editors: M. J. Kennish, R. A. Lutz
1983. XIV, 396 figures. ISBN 3-540-90935-4

Contents: Introduction. – Physical Description of Barnegat Bay. – Nitrogen Distribution in New Coastal Bays. – Phytoplankton. – Macroflora. – Zooplankton. – Benthic Fauna. – Shellfish. – Shipworms. – Fouling Organisms. – Fishes of Barnegat Bay, New Jersey. – Commercial and Sport Fisheries. – Trophic Relationships. – Antropogenic Effects on Aquatic Communities. – Summary and Conclusions. – Appendix A: Bibliography of Unpublished Technical Papers, Theses, and Government Reports on Barnegat Bay, New Jersey. – Author Index. – Organism Index (Scientific Name). – Subject Index.

Volume 7
W. R. Edeson, J.-F. Pulvenis
The Legal Regime of Fisheries in the Caribbean Region
1983. X, 204 pages. ISBN 3-540-12698-8

Contents: Introduction. – The International Law Background. – National Legislation Relating to Fisheries. – Bilateral and Joint-Venture Fisherie Agreements. – Fisheries Administration. – Conclusions. – Tables. – Bibliography.